The Malevolent Designer

Why Nature's God is Not Good

Rosa Rubicondior

Illustrated by
Catherine Hounslow-Webber

The Malevolent Designer

Front cover artwork and all illustrations drawn by

Catherine Hounslow-Webber, except:

Figure 2 View from Wittenham Clumps, Little Wittenham, Oxfordshire – photograph by Catherine Hounslow-Webber

Figures 7 & 8 Covid-19 Cases, USA and UK, Johns Hopkins University

Figure 50 X-ray of child with rickets - Internet source

ISBN-13: 979-8670361729

The Epicurean Paradox

Evil exists.

Is God willing to prevent evil, but not able? Then he is not omnipotent.

Is he able, but not willing? Then he is malevolent.

Is he both able and willing? Then whence cometh evil?

Is he neither able nor willing? Then why call him God?

Attrib. Epicurus (341 – 270 BCE)

I form the light, and create darkness: I make peace, and create evil: I the Lord do all these things.

Isaiah 45:7 – KJV

Prove all things; hold fast that which is good.

1 Thessalonians 5:21 KJV

The Malevolent Designer

Dedication

This book is dedicated to the small army of scientists, rationalists, enthusiastic amateurs and people more interested in the truth than in the smug certainty of easy answers, who daily combat the torrent of misinformation, disinformation and lies that pass for Creationism on the social media, who are often attacked with, condescension, abuse and threats by people with the scientific literacy and foot-stamping, debating tactics of a backward 10 year-old - tactics which will be familiar to readers of my parody of Creationist apologetics in, "*The Internet Creationists' Handbook*" (ISBN-13: 978-1721605149).

It is also dedicated to the readers of my 'Rosa Rubicondior' blog (https://rosarubicondior.blogspot.co.uk).

The Malevolent Designer

Contents

Contents

Introduction

Warning! If you prefer a Panglossian view of the world which sees everything for the best, in the best of all worlds and the delusion that a perfect, benevolent designer designed the best of all possible world just for you to gaze in wonder at, then this book is not for you.

Figure 1 View from Wittenham Clumps, Little Wittenham, Oxfordshire. Photo by Catherine Hounslow-Webber

If you want to look at nature and see nothing but perfection where all things are created for a purpose, and all things work together in perfect harmony, then you won't want to think about the bacteria, viruses and parasitic organisms that are making life difficult for very many of the things you can see, for no other purpose but to make more copies of themselves.

If you want to look at a beautiful butterfly and see nothing but perfection, then you won't want to think about the bacteria living in its cells or the parasitic wasp whose grubs turned its caterpillars into zombies so they could eat their insides to make more parasitic wasps, then this book is not for you. It will spoil your delusions.

If you are of a nervous disposition then you should avoid looking too closely at the natural world – which is full of the stuff of nightmares – and you should avoid this book. If you want to retain the cosy notion of a magic, benevolent sky man looking lovingly and protectively at you, then you should avoid this book. It will shatter that delusion too. Nature's god is not good. Nature's god does not care. Nature's god is supremely indifferent to suffering and to our desire for love and peace. Nature's god is powerless to intervene and turn the world into a friendly place, free from suffering and want. That place exists only in the minds of those who know nothing of the detail of natural world and who are determined to stay that way.

If, however, you want to see the natural world in all its glory, warts and all, and you want to understand how amoral, uncaring, undirected and aimless natural processes created it and why the natural world is the way it is, then this book should help you resist any temptation to see it other than what it is – the glorious result of 3.8 billion years of evolution by natural selection, where not a whiff of magic was involved; an open and accessible world which is amenable to reason without resort to magic or ineffable mysteries forever close to us.

In my previous book in this series, *The Unintelligent Designer: Exposing the Intelligent Design Hoax* (1), I exposed the hoax behind the Intelligent Design movement, showing how the superficial semblance of design in living things quickly disappears when you look below the surface and compare what we see with the principles of good, intelligent design.

Introduction

What we see is massive redundancy at all levels, from junk DNA, to atavistic genes, to vestigial structures and redundant biochemical pathways. This creates huge and unnecessary complexity in all biological systems.

We also see prolific waste on a vast scale, from the giant puffball producing tens of trillions of spores where only one is needed to maintain a stable population to oak trees producing trillions of pollen grains a year to ensure one acorn produces one tree over a lifetime of maybe several hundred years, to the malaria-causing parasite that produced millions of copies of itself to ensure a few get passed to the next generation of victims..

And in the genomes of all multicellular organisms we see the hugely complex system of epigenetics to turn most of the genome off in specialised cells to ensure they stay specialised, and a mechanism for resetting the entire thing in the zygote to allow this specialisation to begin again from a cell with the full potential to become any required speciality – a pluripotent cell. Every one of your approximately 17 trillion cells, save only your red blood cells, have your entire genome replicated within it, yet each uses a mere fraction of the genes to perform its specialised function.

This epigenetic system is only necessary because multicellular organisms have inherited the same cell–reproducing method used by their single–celled ancestors to reproduce entire new organisms, where copying the entire genome was necessary. An intelligent designer could have designed a means of reproducing specialised cells that did not entail copying every DNA codon, with all the attendant risks of mistakes and wasted resources, so only the required genes were replicated. Evolution, being purely utilitarian and having no plan, has produced instead a design for which any intelligent designer would be sacked for gross incompetence.

And we see no clear purpose beyond the mere perpetuation of the species for the sake of it, unless we look at what the organisms actually do and assume this is their purpose – and that brings me on to the subject of compassion and moral intent.

What we also see in nature, when we look beyond the superficial, is a conspicuous absence of compassion; an amoral, indifferent process that shows no signs of anything which could be called benevolence; something I alluded to several time in *The Unintelligent Designer* although strictly speaking this is not an argument against intelligent design *per se*, hence this book. It would be perfectly logical to argue that the putative designer had malevolent intent; that for some reason it designed suffering into the system for some gratuitous, sadistic reason of its own.

Such a designer would, of course, be worthy of utter contempt and revulsion; the moral equivalent of someone who breeds kittens for the sadistic pleasure of burning them alive. This creator could not conceivable be worthy of love and adoration or construed as the source of ethics and moral codes fit for a civilised society. Indeed, anyone who proposed such a thing should probably be removed from civilised society for all our sakes.

Nature is full of wonder and majesty; breathtakingly beautiful; endlessly fascinating and begging to be discovered and admired for what it is. What it is not, is loving, caring and compassionate and an inspiration for human behaviour and a just society.

The reason ID, and Creationism, holds such an appeal to a minority of people was explained recently by a team of psychologists working in France (2) who showed that the cause of Creationism is the same as that of Conspiracism – a childish mode of thinking retained in adulthood to become a disorder – the teleological thinking fallacy.

Children normally think in terms of agency to explain why things happen. Ask a child why it rains and they will usually explain it as something wanting or needing it to rain - because the plants are thirsty, because clouds want to drop their water, because rivers need water, etc.

Introduction

This teleological thinking seems rational to children because nothing important happens in their world without someone intending it to. And they inhabit a world of authority figures.

Creationists retain this teleological thinking fallacy, so their thinking is a simplistic, neotenous process requiring everything to have an explanation in terms of agency and intent. Many of them seem to find it impossible to imagine anything happening by chance or because of a natural process with no intent behind it. I have had Creationists argue that no chemical process can ever happen unless God wants it to; no protein can ever fold in a specific way unless God folds it. God, apparently, is busy folding every one of countless trillions of protein molecules and making every atom react with others to form molecules. All chemistry and physics require the active participation of a magic god to make things happen!

To a Creationist, every one of your 17 trillion cells with its complete complement of DNA and millions of proteins was personally made by God who directed all the chemical and physical processes involved. And not just for you but for each and every living organism - for all of the history of life on Earth and on every other planet where life might exist.

And then, of course, there are all the other physical and chemical processes in the entire Universe.

Asking Creationists to accept that the Big Bang did not have a cause or was not willed to happen, or asking them to think apparent design might not be the result of actual design or that natural selection happens naturally without a plan, is asking them to accept what to them seems ridiculous. It is like asking a child to accept that dinner prepares itself and puts itself on the table or that a Boing 747 spontaneously self-assembles. The distinction is lost on them.

Their thinking, and often their reasoning ability and behaviour, as can be witnessed on the social media, is childishness amounting in many cases to a mental disorder. If they were in a very small minority, their views would be considered eccentric, even psychotic. It is only because they exist within a religious culture where belief in a god is mainstream and 'faith', i.e. belief in the absence of evidence or even despite contradictory evidence, is considered a virtue, that Creationism is merely seen as religious fundamentalism, albeit a cultish one, not a childish delusion.

But then to maintain this delusion, they need to ignore very much of what can be freely seen in nature. The argument from ignorant incredulity can only be maintained by carefully maintaining the ignorance.

Additionally, children brought up by fundamentalist parents were shown recently to be unable to distinguish between fact and fantasy/fiction because religion teaches them to accept magic as reality. Children raised in a secular household have no such disability (3). This, combined with a teleological mind makes the children of fundamentalists susceptible to magic 'explanations' such as creationism, where a magic man supposedly did things by magic and everything magically popped into existence.

In this book, I will show the sheer malevolence of any putative designer, looking at nature from the assumption that such an unpleasant designer actually exists and what it would tell us about it, if that assumption were true.

This, if they ever acknowledge it, represents a serious problem for the Intelligent Design movement because they have no way of explaining how the arms races and parasite–host relationships that abound in nature and explain much of the biodiversity we observe, could be the work of an intelligent designer because it is simply not intelligent to compete with yourself in an arms race, trying at each iteration so solve the 'problem' you just designed as a solution for the other organism involved.

For example, humans are susceptible to infection by bacteria, viruses fungi, parasitic worms, etc., so we have an immune system to help protect us, yet very many would–be parasites have systems for overcoming our immune defences, against which we, in some instances, have developed further defences. What's going on here, that a designer is designing an organism to treat us as a host, then designing an immune system to prevent them doing so, then designing a means to overcome it, and finally, a means to plug that gap in our defences?

One solution would be for the ID movement to conclude that there must be more than one designer at work here; maybe hundreds or thousands of them. Maybe one for each species, even, each one looking after its own design in competition with the others; sometime forming alliances; sometimes becoming mortal enemies. If we conceded design, forgetting for the moment the evidence of a lack of intelligence and an absent ability to plan ahead or to hit the reset button and start again, then why not conceded a multitude of designers?

And why not conclude that the putative designer is indeed a malevolent sadist at worst, or completely indifferent to suffering at best? But of course, a pantheon of creator gods is anathema to fundamentalists of all the Abrahamic, strictly monotheist, religions.

What many Creationists do at his point in the argument is to invoke 'The Fall', where Christian theology states that 'sin' entered the world because Eve tempted Adam to eat an apple (honestly!) and it has been going downhill ever since. This must cause consternation in the ID movement's leadership who invented ID specifically to try to make Creationism look like science, so it could be taught in school science classes at public expense, so circumventing the Establishment Clause of the First Amendment to the US Constitution. The last thing the movement's leaders want to hear is ID being shown to require Christian fundamentalism to make it work. On the other hand, they can't concede multiple designers or malevolent intent in design. So, they have a problem, which is often solved by the time–honoured religious apologetic strategy – ignore it completely. Hence you will rarely, if ever see parasitism being discussed in an ID disinformation website or book.

The reason of course, is that Intelligent Design is nothing more than Bible literalist fundamentalism dressed in a lab coat and made to sound like science to those who are mostly ignorant of much of science. As such, the putative designer also has to conform to the Christian notion of a supreme god which demands that this is the only god; a loving, benevolent, maximally good god with humans as its special creation, living on a specially designed planet on which everything is all about humans. This must be so because this is what fundamentalist religious dogma states, not science, or scientific evidence, but what it says in the Christian Bible.

This book shows why, even if we concede for the sake of argument that a designer god really **is** behind the scenes designing stuff, it can't possibly be described as benevolent or anything approaching maximally good.

Maybe ID advocates, if they are really concerned to promote and defend their god, should consider just how badly it comes off with anything more than a rose-coloured view of nature and a gross distortion of the reality they believe it created. It won't do to pretend the nasty things are nothing to do with a god who reputedly created everything in full knowledge of what it would do.

How is this god best served by having to turn a blind eye to much of what it supposedly does? Why do they need to ignore much of what they believe their loving god created and even regard it as a forgery, or lie intended to mislead as some sort of test? What are ID advocates ashamed of? Or do they believe another creator is also doing some creating – a creator over whom their god is unable to exert any influence or control and who is able to modify its designs and frustrate its intentions at will?

Introduction

In the following pages, I will provide details of natural organisms and relationships that creationists are obliged to attribute to the work of their designer god, all the while asking the question:

Why is the only explanation for this that doesn't leave creationism's god looking like a sadistic, misanthropic, pestilential, megalomaniacal monster, evolution by natural selection?

Lastly, the enjoinder I preface all my books with: never accept anything I say on my authority alone. Check everything. Hopefully, when you do that. you'll find I am right.

The Malevolent Designer

A Little Nastiness

The world of parasitology is a rich source of examples of 'designs' that, if they were the work of a sentient designer could only be the product of malevolent intent. In this chapter, I'll mention just a few.

Goldenrod Soldier Beetles and Fungal Parasites.

The goldenrod soldier beetle, *Chauliognathus pensylvanicus*, is a rather handsome, yellow and black North American long-horn beetle. It eats the flowers of the frost aster, the common boneset and the Canada goldenrod and normally mates on top of these flowers. Its wasp-like colour is an example of defensive or Batesian mimicry, but it is harmless.

Figure 2 Goldenrod Soldier Beetle

It is host to a fungal parasite that for its sheer nastiness would be astonishing if only there weren't so many competitors, each vying for the title of nasty of nasties.

The fungus, *Eryniopsis lampyridarum*, lives inside the body of female beetles, consuming her insides, leaving enough to keep her alive until the fungus is ready. Then it kills her.

Not content with killing her, it then proceeds to reanimate her dead body by spreading her wings – the invitation to any passing male to mate, and so the fungal spores can better escape through the spiracles she had been using to breathe through. Then, in a particularly ghoulish twist, it makes her body swell so that to a male she looks full of eggs and ready to mate.

Any male that attempts to mate with her becomes contaminated with fungal spores on his genitalia, with which he then infects the next female he mates with (4).

An ID proponent is required to believe an intelligent designer designed the soldier beetle and its reproductive system for this fungus to use. As we will continue to see, ID proponents are required to believe their putative designer is capable of many such horrors, while proclaiming it to be all-loving.

Millipedes and Genital Fungi.

One day while perusing some photographs of centipedes sent to her by a colleague, via the social networking site, Twitter, Anna Sofia Reboliera, a biologist with the University of Copenhagen, noticed signs of a fungus never before seen on an American centipede.

The fungus was one of almost 2200 species of fungi from the *Laboulbeniales* family which infects mostly insects, This fungus was on the genitalia of the centipede, *Cambala annulate*! She named it *Troglomyces twitteri* in a tribute to the social medium she learned of it through. The fungi do not look like typical fungi and look more like miniature sponges. They feed by penetrating their host's outer shell. They are almost exclusive to insects and other arthropods such as millipedes. (5)

The benefits to a parasite from infecting the genitalia of its host are obvious, as we saw in the previous example. It virtually guarantees that they'll be passed on to other hosts, but it would take a special sort of designer to create such a design and not one that could conceivably be called loving or benevolent.

Chronic Bee Paralysis Virus (CBPV).

This one is a virus that is spreading quickly in the UK and which kills honey bees, often wiping out the entire hive, as though the *Varroa destructor* mite (of which more later) wasn't doing it quickly enough.

Symptoms of infection include abnormal trembling, loss of the ability to fly and the appearance of a shiny black, hairless abdomen.

By analysing government bee health inspection records, researchers at Newcastle University, UK, showed that the incidence of CBPV had increased exponentially in the UK between 2007 and 2017, and was now present in 39 of England's 47 counties and in 6 of 8 Welsh counties, having only been recorded in Lincolnshire in 2007 (6).

Honey bees are, of course, an important pollinator and are essential to human agriculture. It has even been claimed that extinction of the honey bee would mean human beings would struggle to survive. According to a BBC website "[Honey bees] are critical pollinators: they pollinate 70 of the around 100 crop species that feed 90% of the world. Honey bees are responsible for $30 billion a year in crops" (7).

It is hard to discern the intelligence or benevolence in a designer who designed honey bees, allegedly for the benefit of humans, then, now we have become so dependent on them, designs a virus to help exterminate them, having first devastated them with a parasitic mite.

Bee Mites and Their Viruses.

Varroa destructor is a parasitic mite that lives by sucking the body fluids of honey bees and which spreads very quickly throughout the hive. But that's not the end of the malevolence here.

Bees could possibly tolerate having to feed a few parasitic mites living on them, but what they cannot tolerate is the nasty collection of viruses the mite carries and for which it almost seems to be designed to be the vector. One such virus for example causes a deformity in the bee's wing, so it is unable to fly.

To any biologist, the explanation for this situation is very clear, but no less potentially tragic. Honey bees are the product of co-evolution between the flowering angiosperm plants and a group of flying insects – the Hymenoptera – that began some 200 million years ago, soon after the collapse of the Carboniferous forests.

Mites are the result of an evolutionary process which began a little earlier, back in the Carboniferous era when a group of arthropods began to diversify from their common ancestor with scorpions, into the arachnids - spiders and mites - each evolving to exploit new niches as they opened up. The mites probably started out sap-sucking but quickly turned to sucking the blood and body fluids of the evolving fauna. We will meet more of their descendant later on as they form a major group of parasites.

Eventually, one mite, *Varroa destructor*, evolved to exploit the niche that colonies of bees had unwittingly provided, and, never slow to miss an opportunity, a collection of viruses evolved to exploit this relationship, much the way other parasites have exploited the opportunity that blood-sucking insects have provided. We will meet some more of those soon, too.

Now, to anyone who understands evolution, that is all very straightforward and even infinitely fascinating, but to an ID advocate there is much to explain here. ID advocates insist that all this was done by an intelligence who knew perfectly well what it was doing and what the outcome would be. So, where is the intelligence in going to all that trouble to create honey bees and their relationship with both flowering plants and humans, only to then threaten them with extinction by designing a parasitic mite and supplying it with lethal viruses to inject into the bees?

And where is there anything that could be regarded as benevolent? Probably there is not, but there is in *Varroa destructor* something that can only be regarded as malevolent.

Next, we will meet a parasite that has been 'designed' to overcome resistance in the vector species that is used to transmit it to its victims.

The Black-Legged Tick and Anaplasmosis.

The black-legged tick or deer tick (*Ixodes scapularis*) is the main vector for spreading the *rickettsia* parasite, *Anaplasma phagocytophilum* that causes anaplasmosis in humans.

According to the American government agency, The Center for Disease Control and Prevention (CDC) (8), the signs and symptoms of anaplasmosis are:

Early Illness

Early signs and symptoms (days 1-5) are usually mild or moderate and may include:

- Fever, chills
- Severe headache
- Muscle aches

- Nausea, vomiting, diarrhoea, loss of appetite

Late Illness

Rarely, if treatment is delayed or if there are other medical conditions present, anaplasmosis can cause severe illness. Prompt treatment can reduce your risk of developing severe illness.

Signs and symptoms of severe (late stage) illness can include:

- Respiratory failure
- Bleeding problems
- Organ failure
- Death

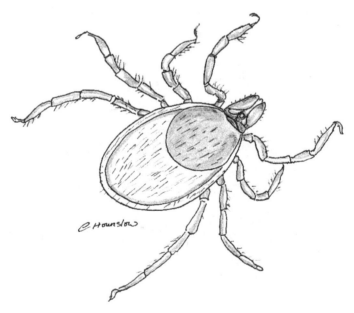

Figure 3 Black-legged tick

If you subscribe to the notion of intelligent design, you might think designing a *rickettsia* parasite to make people sick and possibly die is enough, but, as a research team from the Department of Biological

Sciences, Old Dominion University, Norfolk, Virginia, USA and Department of Entomology, Connecticut Agricultural Experiment Station, New Haven, Connecticut, USA have discovered, that was not the only thing this putative designer designed. It had also designed the parasite to overcome the natural defence of the tick (a defence that the same designer designed to protect the tick in the first place!)

What they discovered was that *A. Phagocytophilum* takes control of a regulatory mechanism that normally limits the amount of substance known as isoatp4056, which is used in transporting organic anions. The more of this substance present in the tick cells, the easier it is for the parasite to make more copies of itself in the cell and the more of them there are available to be injected into the next host when the tick takes a meal (9).

Now, you might be thinking how much further does this putative designer need to go in its endeavours to make us sick. If the next example is anything to go by, there is nothing that it won't stoop to.

Stomach Bugs and Swimming in Gloopy Poop.

Normally, ID advocates love anything to do with bacterial flagella, believing them to be an example of 'irreducible complexity' and therefore evidence for an intelligent designer, no matter that this notion was comprehensively refuted by Kenneth R. Miller, amongst others (10), soon after Creationist, Michael J. Behe invented the idea (11), and has been dismissed in several US court cases as Creationism trying to look like science to fool legislators (12). So, they should relish the discovery that one of the bacteria that causes food poisoning, *Campylobacter jejuni,* uses an ingenious mechanism of two flagella to help it overcome the problem of being 'designed' to live in the thick, sticky contents of the human lower digestive system.

Researchers at Imperial College, London, Gakushuin University, Tokyo and the University of Texas Southwestern Medical Center, found that *C. jejuni* actually swims **faster** in thick, sticky liquid, not more slowly as you might expect, so they filmed a strain that has been genetically modified to have fluorescent flagella, using high-speed photography, to work out how it achieved this. It seemed strange that a bacterium has a flagellum at each end, rather like a boat with two opposing outboard motors.

What the researchers saw was that the bacterium wraps the leading flagellum around its helical shaped body so both worked to drive it in the same direction. To reverse, they simply change the flagellum which is wrapped round their body. And it turned out that wrapping the flagellum around the body was easier in thick stick liquid. In thinner, less viscous liquids, the bacterium was unable to wrap its leading flagellum around itself, so motility became more difficult (13).

We are supposed to believe that an intelligent designer went to all the trouble to design an organism to give us food poisoning then put it in an environment that needed a special, complex mechanism to solve the problem of motility it has just created – and all to make us sick. Then it bizarrely placed two of its intelligently designed, 'irreducibly complex' (according to Michael J Behe) flagella, not at the same end but at opposite ends, so needing this bizarre and clunky, complex work-around system so it could move about and do its job – giving us food-poisoning – at least, given that apart from making more copies of itself, that is what it does, so we have to assume any intelligent designer designed it for that purpose.

Meningococci and Septicaemia.

Neisseria meningitidis also called *Meningococcus* is a pathogenic bacteria which can cause meningitis and septicaemia. Scientists working at Würzburg University, Bavaria, Germany recently discovered that it is assisted in this by a small protein known as ProQ.

ProQ among other things is involved in DNA repair and resistance to oxidative stress. Both these increase the virulence and pathogenesis of the bacteria but where ProQ really has an impact is in activating up to 250 bacterial genes that are involved in the motility and pathogenic effects of infection (14).

Whatever 'designed' this nasty little bacterium didn't leave much to chance in its determination to make people sick and die – assuming, as ID proponents do, that it **was** designed of course.

Incidentally, so pleased was the designer with the success of this protein that it also gave it to two other nasty little pathogens - *Escherichia coli* and *Salmonella enterica* (15) and the virulence of *Salmonella* actually depends on it (16).

The irony of *E. coli* having this protein, when the flagellum of *E. coli* was the subject of Michael J. Behe's book (11) with which he introduced the notion of 'irreducible complexity' (which he asserted was evidence for a designer). His book, which was greeted with derision and disdain by biologists, many of whom pointed out the factual errors in it, and rapturous enthusiasm by Creationists, is now almost a sacred text to the ID movement. He was also, probably unwittingly, introducing the world to the idea of a malevolent designer.

Male-Killing *Wolbachia.*

The *Wolbachia* group of bacteria are common parasites, symbionts or commensal bacteria on a whole range of insects and other arthropods. They are believed to infect some seventy percent of insects. One such insect it the tsetse fly, responsible for spreading the *Trypanosoma* protozoans that cause sleeping sickness, where *Wolbachia* may be essential for the tsetse fly to reproduce.

One insect in which *Wolbachia* is parasitic is the pretty butterfly known as the common eggfly, *Hypolimnas bolina*. *Wolbachia* is normally passed on to the next generation in the eggs the females lay so, to the bacterium, males have no purpose and compete with females for resources, so it kills them. 'Male' eggs infected with Wolbachia normally fail to hatch, as the caterpillar dies in the egg. Females are unharmed. In some species, *Wolbachia* 'converts' male embryos to females. Observations have shown that female eggfly butterflies become more promiscuous as the ratio of males in the population diminishes so partially offsetting any problems finding a mate that *Wolbachia's* male-killing strategy might cause them. Killing the males ensures more resources are available for the females, so benefiting the *Wolbachia*.

In a 2001 survey of the Samoan Islands of Upolu and Savai'i, predation by *Wolbachia* was found to have reduced the proportion of males to only about one percent of the population. When the islands were again surveyed during 2005, it was found that on Upolu the male to female ratio was 1:1 but on Savai'i out of 100 specimens caught, not a single male was found. The males on Upolu had acquired a gene which suppressed *Wolbachia*. But, and this is where the rate of evolution was most astonishing, by 2006, the ratio on Savai'i was also approaching 1:1. In a single year the suppressor gene had reached Savai'i and males were increasing rapidly in the population. These males were also carriers of the mutant, suppressor gene. Hatching success rates of the eggs had also improved in line with this recovery. The fact that just about every male carried the mutation, having started from a position where there were very few males, meant that every successful mating increased the proportion of males in the population. (17).

The standard scientific explanation is that this is a case of evolution by natural selection, first to evolve Wolbachia so it kills males because this favoured *Wolbachia*, then to change *H. bolina* so the males have immunity, because this mutation favoured male *H. bolina*. Presumably, any ID 'explanation', if there is one, would involve the omniscient designer changing its inerrant mind and deciding not to exterminate the local populations of *H. bolina* after all.

Methicillin Resistant *Staphylococcus aureus.*

N. meningitidis, E. Coli and *S. enterica* are not the only bacteria to have received the attention of Creationism's putative intelligent designer to help them resist the efforts of human medical science. *Staphylococcus aureus* has also been 'intelligently' modified to make it resistant to antibiotics, so Methicillin Resistant *S. aureus* or Multi-resistant *S. aureus* (MRSA) is now a major problem in hospitals.

As I was writing this book, news came that a team of scientists from the University of Sheffield have discovered the mutations that give *S. aureus* this resistance without reducing its pathogenic effects. (18) It is due to changes in the genes which code for the protein enzyme RNA polymerase. In other words, according to the ID view, *S. aureus* has been redesigned to make it just as harmful while being resistant to our defence against it, in what can only be described as an arms race between science and the bacterium's designer.

Given their dogmatic refusal to acknowledge that the process of evolution by natural selection could account for advantageous mutations increasing in the *S. aureus* gene pool, ID advocates have no choice but to assign the result of this arms race to a sentient designer.

Salmonella and Antibiotic resistance.

Salmonella comes in two related species: *Salmonella enterica* and *S. bongori*. Both can cause the condition known as Salmonellosis. *S. enterica* has 6 subspecies. The Center for Disease Control and Prevention (CDC) lists the symptoms of Salmonellosis as:

Most people with *Salmonella* infection have diarrhea, fever, and stomach cramps.

Symptoms usually begin six hours to six days after infection and last four to seven days. However, some people do not develop symptoms for several weeks after infection and others experience symptoms for several weeks.

Salmonella strains sometimes cause infection in urine, blood, bones, joints, or the nervous system (spinal fluid and brain), and can cause severe disease (19).

Usually, *Salmonella* remains within the intestinal tract but occasionally it can enter the bloodstream where it can cause the life-threatening, paratyphoid fever, which needs to be treated with antibiotics.

Again, according to the CDC, *Salmonella* cause about 1.35 million illnesses, 26,500 hospitalizations, and 420 deaths in the United States every year. People get Salmonellosis from:

- Eating contaminated food or drinking contaminated water

- Touching infected animals, their feces, or their environment

Those most at risk from *Salmonella* are:

- Children under 5 years old are the most likely to get a *Salmonella* infection.

- Infants (children younger than 12 months) who are not breast fed are more likely to get a *Salmonella* infection.

- Infants, adults aged 65 and older, and people with a weakened immune system are the most likely to have severe infections.

- People taking certain medicines (for example, stomach acid reducers) are at increased risk of infection. (19)

Salmonella seems to be the organism of choice for Creationism's intelligent designer to design antibiotic resistance in its arms race with human medical science. Researchers at Cornell University's Department of Food Science recently discovered that *Salmonella* had developed resistance to the antibiotic Colistin, described by the World Health Organization (WHO) as the 'antibiotic of last resort'.
Salmonella has the mcr-9 gene, the 9th in a line of antibiotic resistance genes, each one a mutation of the previous one, and each one improving antibiotic resistance. This gene can be freely passed to other bacteria in the form of a plasmid (a small section of DNA than can be passed from one bacterium to another, even another species (20).

Imagine going to all that trouble just to make people sick!

Toxoplasma Takes Control.

Toxoplasma gondii is a parasitic organism that causes the disease, Toxoplasmosis. You can accidentally catch it from your cat, through its faeces.

According to The American Center for Disease Control and Prevention (CDC):

Healthy people (nonpregnant)

Healthy people who become infected with *Toxoplasma gondii* often do not have symptoms because their immune system usually keeps the parasite from causing illness. When illness occurs, it is usually mild with "flu-like" symptoms (e.g., tender lymph nodes, muscle aches, etc.) that last for weeks to months and then go away. **However, the parasite remains in the person's body in an inactive state. It can become reactivated if the person becomes immunosuppressed. [My emphasis]**

Mother-to-child (congenital)

Generally, if a woman has been infected before becoming pregnant, the unborn child will be protected because the mother has developed immunity. If a woman becomes newly infected with *Toxoplasma* during or just before pregnancy, she can pass the infection to her unborn baby (congenital transmission). The damage to the unborn child is often more severe the earlier in pregnancy the transmission occurs. Potential results can be

- A miscarriage
- A stillborn child
- A child born with signs of congenital toxoplasmosis (e.g., abnormal enlargement or smallness of the head)

Infants infected before birth often show no symptoms at birth but may develop them later in life with potential vision loss, mental disability, and seizures.

Persons with ocular disease

Eye disease (most frequently retinochoroiditis) from *Toxoplasma* infection can result from congenital infection or infection after birth by any of the modes of transmission discussed on the epidemiology and risk factors page. Eye lesions from congenital infection are often not identified at birth but occur in 20-80% of congenitally-infected persons by adulthood. However, in the U.S. <2% of persons infected after birth develop eye lesions. Eye infection leads to an acute inflammatory lesion of the retina, which resolves leaving retinochoroidal scarring. Symptoms of ocular disease include

- Eye pain
- Sensitivity to light (photophobia)
- Tearing of the eyes
- Blurred vision

The eye disease can reactivate months or years later, each time causing more damage to the retina. If the central structures of the retina are involved there will be a progressive loss of vision that can lead to blindness (21).

As well as causing a whole range of debilitating and disabling conditions, *T. gondii* has the ability to remain dormant in the body until its host becomes sick through some other cause, then, to add insult to injury, it will reactivate.

Now scientists from Indiana University's Department of Pharmacology and Toxicology have discovered that *T. gondii*, which infects up to one third of the human population, gains control of its host's cells, causing them to migrate to other parts of the body (22). In other words, by a mechanism which is so far unknown, *T. gondii* hijacks the body's immune system and uses it to spread itself to other organs where it can remain dormant until reactivated.

This is not the only way *T. gondii* manipulates its host either. There is some evidence that it induces personality changes in its human hosts (23). It has been shown to make rats attracted to cats (24), for example, and in what could be relevant to humans, infected chimpanzees (*Pan troglodytes troglodytes*) have been shown to have a morbid attraction to the scent of leopard (*Panthera pardus*) urine (25). It could be that the personality changes associated with *T. gondii* infection in humans is an echo of its manipulation of a common ancestor we share with the chimpanzees. *T. gondii* manipulates its hosts to get them eaten by a predator so it can infect that predator!

You might well be asking yourself now just whose side the designer of *T. gondii* is on here. Having, supposedly, designed the mammalian immune system to protect mammals against infectious organisms, it then appears to have designed a work-around for what it sees as a problem from T. gondii perspective.

It doesn't look anything like a designer who only wants the best for its favourite creation. In fact, it looks like a designer who delights in inventing ever-more ingenious ways to make life difficult for humans and to facilitate the harm parasites do.

A similar thing can be seen in our next example.

Influenza Virus H3N2 Beating Medical Science

A problem a virus like the common H3N2 influenza virus faces is that its hosts have an immune system that makes anti-bodies that recognise and kill it, so preventing it re-infecting a person who has already had it. When enough people have had it, so-called herd immunity builds up in the population, so, even if one person catches it for the first time, the chance of it being passed on to the next person is greatly reduced. Eventually the virus' environment becomes too difficult for it to thrive in, so, unless it mutates to a form the immune system doesn't recognise, it will die out. One might be forgiven for thinking this system looks designed to produce a never-diminishing supply of novel viruses because that is exactly the result of it.

What human medical scientist have developed is a system of vaccinations which produce antibodies in people before they catch the virus, so building this herd immunity very quickly and protecting especially vulnerable people.

However, a team of researchers from the Johns Hopkins Bloomberg School of Public Health recently discovered is that the H3N2 virus has acquired (or as an ID advocates would say, has been redesigned to have) a mutation which makes **it** immune to human antibodies, by preventing them from attaching to the proteins in the viral coat that the virus uses to stick itself to the cell wall and gain entry into the cell. The same mutation appears to make the virus good at replicating inside a cell once it gains entry.

Because different strains of the same virus can infect the same person simultaneously and because the modified virus is good at using the available resources in the host cell, this mutation is now present in almost all strains of the H3N2 virus and humans have a reduced ability to defend themselves from it (26).

With a single mutation, H3N2 has circumvented both our natural and artificially-induced defence.

From the point of view of an evolutionary biologist, this makes perfect sense as the mindless and purposeless mutation gives the virus greater fitness in its environment, so it makes more copies of itself, complete with the mutant gene. From the point of view of an ID advocate, however, there is no way this can be presented as either intelligent or benevolent. Their 'designer' appears to have regarded the immune system it reputedly designed, and the efforts of medical science it reputedly gave us the intelligence to develop, as problems to be overcome, so its viruses can continue to do what they seem to have been designed to do – make us sick.

Streptococcus and Dental Caries.

Dental caries are areas where the enamel coats of our teeth have been eroded by bacteria. Unless treated, they enlarge and deepen and eventually erode through to the centre of the tooth so bacteria can gain access and set up painful abscesses in our jaw which can lead to generalised septicaemia and even death.

Recently scientists from the University of Pennsylvania School of Dental Medicine and the Georgia Institute of Technology showed how they do this and help evade our normal dental hygiene precaution of regular brushing. Essentially, they do it by working together, organised by *Streptococcus mutans* to form a protective shell around the colony which is working away at the enamel.

The researchers found that *S. mutans* is present in dental plaque in discrete mounds against the tooth surface. These mounds were organised with *S. mutans* on the inside covered by other bacteria, such as *S. oralis*, which formed additional outer layers precisely arranged in a crownlike structure. Supporting and separating these layers was an extracellular scaffold made of sugars produced by *S. mutans*, effectively encasing and protecting the disease-causing bacteria. Tests also showed that *S. mutans* was protected from antibiotics by this protective mound (27).

To add to ID advocate's discomfort, it is a central dogma of Creationism that complex organisation cannot arise naturally and must be the result of intentional design. Yet here we have an example of bacteria apparently being designed to work together to create an organised, complex structure, for no other purpose than to make copies of themselves and to give us toothache and worse.

On the subject of dental hygiene, while this book was in preparation a letter was published in the British Medical Journal (BMJ) confirming earlier findings by team of researchers from Harvard's T.H. Chan School of Public Health, Boston, USA that showed a statistical link between gum disease caused by bacteria and some forms of cancer. People who have lost a tooth due to gum disease have a significantly increased risk of developing oesophageal and gastric cancer. (28). The letter suggested a mechanism – *Porphyromonas gingivalis* reduces the immune response of T-cells, reduces apoptosis – the normal, controlled death of cells - dehydrogenation of ethanol to acetaldehyde, causing DNA damage, mutation, and excessive proliferation of epithelial cells. (29)

The human mouth plays host to a number of organisms, some of which are parasitic. Amongst them are a so-called 'red complex' of bacteria, which includes *P. gingivalis*, *Treponema denticola*, and *Tannerella forsythia* which work together to create gingivitis – a disease which, if left untreated, can lead to the destruction of the structures which hold the teeth firmly in place. This can lead to teeth becoming loose and eventually falling out (30). Now it seems at least two of these, *P. gingivalis* and *T. forsythia* may be responsible for producing 'endogenous nitrosamines' which are known to cause gastric cancers.

So, according to ID advocates, their intelligent designer has designed a complex of parasitic bacteria to live in our mouths and cause our teeth to loosen and fall out. Not content with that, it designed them to give us a significantly-increased risk of a couple of rather nasty cancers too.

Just a few more examples should suffice to demonstrate that the only explanation for a great deal of biology, especially parasite biology, which does not leave any putative designer looking like an incompetent, malevolent designer, is evolution. It is therefore a mystery why fundamentalists of so many religions, prefer the malevolent incompetence explanation to the scientific one – evolution by natural selection.

Malaria Parasite's Internal Clock.

Plasmodium falciparum is the parasite that causes malaria, accounting for millions of deaths world-wide. I cover it extensively in "*The Unintelligent Designer: Refuting the Intelligent Design Hoax*" as an example of both needless complexity and prolific waste. Here I will look at how a putative designer is making it better at making people sick and die.

As part of the lifecycle of the parasite, there is a periodic simultaneous release of parasites from the red blood cells of its host. This synchronized release ensures any blood meal taken by a female mosquito will contain plenty of *P. falciparum* organisms to ensure they are successfully passed on to the next victim. It also raised the hosts body temperature, making them more easily detected by mosquitoes.

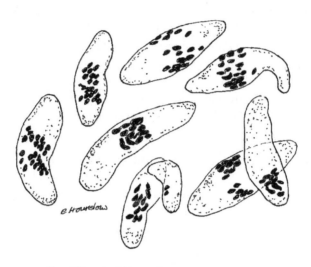

Figure 4 Plasmodium falciparum. (From a stained microscope slide).

The question was, how is this synchronicity achieved.

This was answered by a team of researchers, led by professor Steven Haase of the Department of Biology, Duke University, Durham, NC, USA and colleagues from Walter Reed Army Institute of Research, Florida Atlantic University and Montana State University. They found that the parasites have intrinsic oscillators associated with circadian rhythms and cell cycles.

Moreover, each different strain of parasites has a different periodicity so helping to keep them separate in the host's blood so mosquitoes will only take up one strain or another. It also seems that the oscillator may work differently to those found in other rhythmic cells which rely on genes switching on to produce proteins which, when they reach a threshold concentration, switch off the genes, in a classical bio-feedback loop. However, the precise mechanism in *P. falciparum* is still unknown.

The putative designer has apparently re-invented the wheel (figuratively, if not literally) and has done so to make *P. falciparum* better at killing people and leaving millions of others severely debilitated.

Very recently, researchers from Duke University, Durham, N.C., USA and the Massachusetts Institute of Technology have shown how *P. falciparum* has been cleverly modified to be able to withstand the raised body temperature in its victims. Every few days a malaria victim's body temperature will rise to a debilitating 105 degrees Celsius or more, yet the parasites are able to carry on eating red blood cells with impunity. As they eat the red blood cells, the ingested fragments are taken into tiny vacuoles within the parasite's cell where they are broken down and digested by digestive enzymes. The vacuoles act like tiny guts.

What a raised body temperature would do is to cause the walls of these vacuoles to break down, releasing their enzymes into the parasite cell and causing it to digest itself. The researchers found that *P. falciparum* protects itself from this by producing a protein called phosphatidylinositol 3-phosphate, or PI(3)P for short. This attracts another protein, called Hsp70, the so-called heat-shock protein, to stick to it and together these reinforce the walls of the vacuoles (31). ID advocates must attribute this to their putative intelligent designer.

And this is not the only thing it has achieved, apparently. It has also modified them to make them resistant to the anti-malarial drugs that looked like they might finally eradicate this nasty little parasite. The parasites have acquired several mutations which work together to act as a pump to remove the anti-malaria drug, chloroquine, from the cell, rendering it ineffective.

Although his claim was refuted comprehensively again by Kenneth R. Miller (32), nevertheless Creationist and Discovery Institute associate, Michael J. Behe rushed out a book (33) in which he claimed, using flawed maths and statistics, that this system could only have been intelligently designed. His basic error was in assuming all the mutations needed had to occur simultaneously in a single cell, so the odds against it being spontaneous were astronomical. This is known as the 'big scary numbers' tactic. It is used extensively to fool scientifically illiterate people and always involves a misuse of statistical methods.

Unwittingly or otherwise, Behe again tied his putative designer to the idea of malevolent intent, since in effect he was arguing that it actively and intentionally worked to circumvent medical science's attempt to eradicate malaria. As Miller showed, an accumulative evolutionary process in a large population is more than capable of producing the mutations required in the same cell, so they don't need to arise simultaneously in the same cell as a single event, as Behe had assumed in his calculations.

The Amoeba that Destroys Your Brain.

The following is based in part on a blog post I wrote in 2015 (34).

Naegleria fowleri is an amoeba which destroys the human brain, but it doesn't do it simply by eating it - that would be too simple for our malevolent designer. It does it by making **us** destroy our own brain, not just because our immune system is not fit for purpose but also because of the way our brain is fitted into our skull.

N. fowleri lives in warm, freshwater pools, normally living on a diet of local bacteria. However, should someone take a dip in one of these pools and gets water up their nose, some of the amoebas might penetrate the mucous membrane and find their way up the nerves to the brain, where they start eating brain cells, causing amoebic encephalitis.

Thankfully, this is a reasonably rare event because it is almost always fatal. Of 132 people in the USA known to have been infected since 1962, only three survived. Infection is more common elsewhere: in Pakistan some 20 people a year die from infection by *N. fowleri*. But the amoeba itself is almost certainly not what actually kills people.

According to a paper published in *Acta Tropica* by Abdul Mannan Baig from Aga Khan University in Karachi, Pakistan, the main culprit could be the host's immune system itself which does most of the damage (35).

The immune system's response is to flood the brain with immune cells. Not only do the enzymes released by the immune cells damage the brain cells themselves, but this response causes inflammation and swelling. Normally, swelling and inflammation around an infection site are exactly what's needed because this brings more blood and more lymphocytes to the area, helping to fight the infection and promote healing. In the brain, however, this response can be disastrous.

The problem is a piece of really bad 'design'. The brain is contained in a bony case which can't expand and from which there is only one significant outlet through which pressure build-up can be dissipated - the foramen magnum, i.e., the hole through which the brain stem passes to form the spinal cord. Basically, when the brain swells, it is like trying to squeeze toothpaste out of the nozzle. The result is compression of the brain stem by a process the medical profession call 'coning'. And this is where the second piece of bad design comes in.

All the deep centres needed to maintain basic life-support such as respiration, blood pressure and heart rate are located in the brainstem and get knocked out by coning because the pressure closes the supplying blood vessels.

Abdul Mannan Baig, of the Department of Biological and Biomedical Sciences at Karachi University, believes he has shown that brain cells actually survive longer without an immune response anyway, so he recommends a treatment for amoebic encephalitis which amounts to over-riding the 'Intelligent Design' of the body and suppressing the body's immune system **before** hitting the parasites with specific drugs. This should help reduce the risks from brain-swelling too.

N. fowleri is not the only amoeba to cause amoebic encephalitis. *Balamuthia mandrillaris* and amoebas of the *Acanthamoeba* genus also do the same.

So, and this should really embarrass Creationists if they knew about it and understood it, medical science is recommending switching **off** one badly designed system to reduce the risks coming from another example of bad design in the human body, all to overcome the harm being done by a nasty little parasite that, if we believe Creationists, was also designed by their putative intelligent designer and, apparently, with the omniscient intention of causing this problem in the first place.

Behaviour-Controlling Rabies

Rabies, caused by the Lyssaviruses, which also cause Lassa fever, infects all mammals and controls their behaviour in ways which are conducive to the spread of the virus to other mammals. It consists of a single strand of RNA wrapped in a protein coat. It infects vampire bats which can pass it on in their bite when taking a blood meal or by spreading it in airborne droplets.

The virus makes its victim more aggressive and paranoid while making swallowing difficult. This encourages biting and the retention of saliva in the mouth so a bite will be more infective. The virus travels along nerves until it reaches the brain. In humans, it is almost always fatal once symptoms appear, and it can take from a few days to a year for symptoms to develop following the bite or scratch of an infected dog, cat or vampire bat.

The malevolent intent of any sentient designer who could design something like the Lyssaviruses and rabies is hard to imagine, yet supposedly devoted followers of it insist it must have designed these things and knew exactly what they would do when it did so.

Controlling *Chlamydia*

Chlamydia is the commonest sexually-transmitted infection worldwide and is caused by one of the *Chlamydiaceae* family of bacteria. These are obligate intra-cellular parasites and are the leading cause of infectious blindness. The commonest species in humans are *Chlamydia trachomatis*, *Ch. pneumoniae*, *Ch. abortus* and *Ch. Psittaci*. *Ch. Abortus* can cause abortion and foetal death in humans and other animals. The NHS list the following symptoms of chlamydia:

Symptoms of chlamydia

Most people with chlamydia do not notice any symptoms and do not know they have it.

If you do develop symptoms, you may experience:

- pain when peeing

- unusual discharge from the vagina, penis or bottom

- in women, pain in the tummy, bleeding after sex and bleeding between periods

- in men, pain and swelling in the testicles (36)

The same NHS website list the following ways in which chlamydia can be transmitted:

How do you get chlamydia?

Chlamydia is a bacterial infection. The bacteria are usually spread through sex or contact with infected genital fluids (semen or vaginal fluid).

You can get chlamydia through:

- unprotected vaginal, anal or oral sex

- sharing sex toys that are not washed or covered with a new condom each time they're used

- your genitals coming into contact with your partner's genitals – this means you can get chlamydia from someone even if there's no penetration, orgasm or ejaculation

- infected semen or vaginal fluid getting into your eye

It can also be passed by a pregnant woman to her baby (36).

The bacterium appears to share some genetic characteristics with cyanobacteria and plant chloroplasts which are believed to have evolved from endosymbiotic cyanobacteria. The fact that the *Chlamydiaceae* are obligate intra-cellular parasites could be a clue to the origins of chloroplasts.

Scientists, led by Dr. Karthika Rajeeve, from Julius-Maximilians-Universität Würzburg, Germany, have very recently discovered that when *chlamydia* enters a cell it reprograms the cell to increase the importation of the amino acid, glutamine. It uses lots of glutamine to make the ring-shaped molecule, peptidoglycan, which is a building block of the bacterial cell wall (37). The importance of this research is that, if the glutamine-importing mechanism is disrupted, the bacterium fails to replicate.

To an ID advocate, it is more evidence of the inventive genius with which their putative intelligent designer is making us sick – presumably to punish us for some assumed transgression. And apparently, this intelligent designer has designed one of the *Chlamydiaceae* to produce abortions!

Zikavirus Special for USA.

Zikavirus is a virus of the Flaviviridae family which originated in Africa but which has managed to cross the Atlantic into South America where it has become established. It is spread by *Anopheles* mosquitoes when a female takes a blood meal, in much the same way as malaria and yellow fever.

The American Center for Disease Control and Prevention (CDC) lists the health effects and risks as:

- Zika virus disease is generally mild, and severe disease requiring hospitalization and deaths are uncommon.
- Zika infection during pregnancy can cause serious birth defects and is associated with other pregnancy problems.
- Rarely, Zika may cause Guillain-Barré syndrome, an uncommon sickness of the nervous system in which a person's own immune system damages the nerve cells, causing muscle weakness, and sometimes, paralysis.
- Very rarely, Zika may cause severe disease affecting the brain, causing swelling of the brain or spinal cord or a blood disorder which can result in bleeding, bruising or slow blood clotting (38).

The major risk is obviously to the foetus during pregnancy as infection can pass across the placenta, resulting in under-development of the brain, leading to microcephaly and mental retardation. ID advocates would have us believe that this virus was intelligently designed so we can suppose that this damage to unborn babies was an intended consequence of infection by Zikavirus.

While this book was in preparation, a team of scientists at the University of California - Riverside published a paper which showed how the Zikavirus has been 'designed' to suppress the host's immune system as soon as it gains access to a cell (39). The host's normal response to a virus attack it to produce type I interferon, but the virus produces a non-structural protein, NS5, which interacts with a key protein, STAT2, and degrades it, so inhibiting the production of interferon. An ID advocate would presumably claim this shows how thorough the intelligent designer was in designing Zikavirus, leaving nothing to chance in its determination to get it past the immune system it designed to protect us from the viruses it designed. Just another example of the sort of arms race with itself that this supposedly intelligent designer indulges in to frequently.

And another team of researcher, led by Leah C. Katzelnick at the University of California, Berkeley, has found that the designer has been even more sneaky and underhand. They found that people who have had the Zikavirus are more prone to the closely-related but much more serious dengue fever (40) – a disease carried by the same mosquitos that carry Zikavirus. At least, ID advocates are obliged to attribute this to deliberate design.

But it gets worse!

Scientists from the National Institute of Infectious Diseases, Shinjuku, Tokyo, Japan (41), have discovered that the Zikavirus has been slightly modified into three sub-types each with different degrees of virulence. The most virulent type is the one moving up from South America into the southern United States. It looks, from an ID advocate's perspective as though their putative designer has made a special effort for the American victims of its virus.

Exterminating Frogs with a Fungus.

Most of the examples I've talked about so far have been organisms and viruses that affect humans, but we are far from being the only species that Creationism's putative intelligent designer seems to have taken an intense dislike to. For example, the world's frogs and other amphibians are currently being decimated by chytridiomycosis, caused by a couple of related Chytrid fungi, *Batrachochytrium dendrobatidis* and *B.*

salamandrivorans. It has been estimated that over 500 different species have been severely reduced in number by this fungal plague, with over 90 extinctions.

These fungi seem to have originated in an area of Southeast Asia by modification of a common, harmless, soil fungus. In that part of the world, the local population of amphibians seems to be resistant to the pathogenic forms of the fungi, suggesting that these fungi frequently become pathogenic and the local population have built up resistance to it.

According to research carried out by a team from the Fenner School of Environment and Society, Australian National University, ACT, Australia, it was resistance in the local population which probably kept the disease from spreading more widely, until human agency intervened to change the environment. They have related the increased trade in amphibian species to the spread of the fungi all over the world where they found species with no evolved resistance (42).

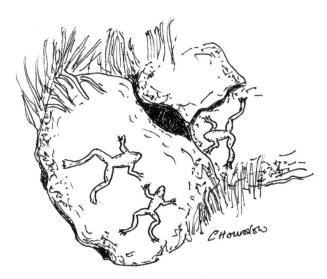

Figure 5 Frog Victims of Chytrid Fungus

Recently, another team found that one of the factors that could have made these fungi so successful is that the frog's immune response seems to have worked against it. Researchers from the University of Central Florida and the Smithsonian Conservation Biology Institute (SCBI) found that, in the frog *Rana yavapaiensis,* a species known to vary in its ability to survive attack by these fungi, those which showed an elevated immune response had a worse outcome that those with a lower response (43). Somehow, the frog's 'designed' immune system was working against it and the fungi had been 'designed' to exploit this.

ID advocates would have us believe that, for reasons unknown, their putative intelligent designer has deliberately redesigned a soil fungus so it can overcome the immune system it designed to protect frogs from infections, and so exterminate over 90 species of amphibians that it designed earlier and severely endanger some 500 species in what has been described as the biggest single loss of biodiversity, albeit, aided and abetted by humans in this endeavour. Creationism's intelligent designer must really hate the frogs it designed. Maybe a private definition of the word 'intelligent' is being employed here.

Yes! We Have No Bananas, Thanks to Fusarium Wilt.

The soil-borne fungus *Fusarium oxysporum* f. sp. *Cubense* is a specialised form of *F. oxysporum* which parasitized banana plants (*Mussa sp.*), causing Panama disease or Fusarium wilt. Entire plantations can be wiped out in a year.

It is believed to have originated in Southeast Asia and was first reported in Australia in 1876. Since then it has spread to almost all banana-producing parts of the world. One method of spread is by infected rhizomes which appear normal. Cultivated bananas are seedless so need to be reproduced asexually with offshoots or rhizomes.

The fungus kills the plant by triggering a self-defence mechanism that produced first a gel, then tylose which block the conductive tissues and prevent water and nutrients from travelling up from the roots.

By the mid-20th Century, resistant cultivars of the 'Cavendish' banana had been developed. However, the fungus has managed to overcome this resistance in the Eastern Hemisphere and this is expected to spread to the Western Hemisphere. Currently, there are no other resistant cultivars available. There is now a real possibility that this economically and nutritionally important crop could become extinct!

Creationism's putative intelligent designer appears to have a down on bananas and has twice modified a soil-born fungus to kill them. Firstly before 1876 and then again in the 20th Century, to overcome resistance in the 'Cavendish' cultivar. It doesn't just stop at bananas, either. *F. oxysporum* also affects the plantain, *Mussa acuminata* × *balbisiana*, which is a staple crop in tropical regions.

Zombie Cicadas.

The 13 and 17-year cicadas of the *Magicicada* genus, so called because of the periodicity of their emergence from an underground larval existence to a mature, reproducing adult, can become infected with a fungus of the *Massospora* genus, such as *Massospora cicadina* (44). This fungus synchronises its production of reproductive spore to coincide with the emergence of these cicadas when it forms a mass of spore-bearing structures to replace the end of the male's abdomen which falls off, making it sterile and eventually killing it.

Males infected with the fungus are induced to mimic female mating behaviour – flicking their wings in response to the mating calls of males to invite mating. They will also tolerate mounting by another male. A male which attempt to mate with an infected male then becomes infected and passes the fungus on to females with whom it mates successfully. The strategy is to infect as many adult cicadas as possible.

Infected cicadas of both sexes spend time just walk around, dragging their abdomen to spread spores in the soil.

Infected adults are the source of spores which are scattered widely in the environment to lie dormant in the soil where they infect the next generation as the nymphs emerge 13 or 17 years later. The fungus appears to coordinate its maturation into the infective stage with the period of the periodic cicada.

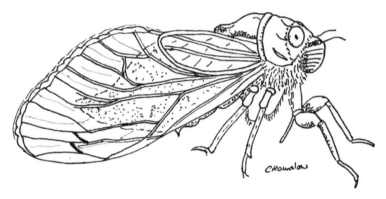

Figure 6 Cicada

There are about a dozen species of *Massospora* each of which is specific to a species of cicada. Studies have shown that when in the body of its host, these fungi produce psychoactive compounds, amphetamine, cathinone, psilocin and psilocybin to manipulate the insect's behaviour (45).

Coronavirus SARS-CoV-2 (Covid-19).

No book dealing with the malevolence of any putative designer would be complete without mentioning the coronavirus, SARS-CoV-2, which causes the acute respiratory illness, Covid-19, especially one written in the midst of a major deadly pandemic not seen since the so-called Spanish flu of 1918-19.

In an effort to control it and prevent health services becoming overwhelmed, government across the world have taken drastic actions, including a complete lockdown, permitting only essential 'key-workers' to leave their homes. Schools, all shops except food shops and chemists, bars, places of worship and restaurants have all been closed and all social gatherings have been banned. This has, in turn, caused economies to collapse and government borrowing to escalate massively.

At the time of writing, many countries, including the UK, are experiencing a resurgence of new infections at least as high that that experienced in the first wave and in many cases, considerably higher. Control measures, which had been relaxed in the summer are being tightened again. Social distancing and wearing face-masks inside public spaces have become the norm. Restrictions have been imposed governing how many people may gather together indoors and out and how families from different households may mix,

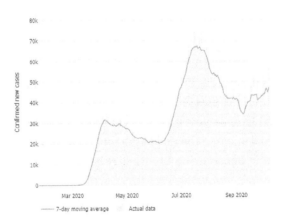

Figure 7 Covid-19 Cases, USA 9 Oct, 2020 (Source: Johns Hopkins University)

As this book went to press, Johns Hopkins University Corona Virus Resource Centre (46) was reporting a total of 36,574,082 cases worldwide, with 1,062,658 deaths. Of these cases, 7,607,950 have been in the USA (212,789 deaths) and 564,502 in the UK (42,682 deaths). US President, Donald Trump, who had been widely criticised for playing down the risks and holding election rallies and events with no social distancing measures or face-masks, is himself recovering from the effects of the virus for which he needed medical intervention in hospital. He draws a great deal of support from the Evangelical wing of fundamentalist Christianity, most of whom are Creationists.

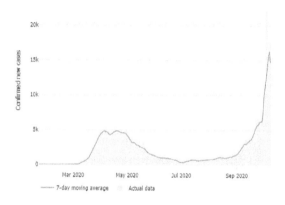

Figure 8 Covid-19 Cases, UK, 9 Oct, 2020 (Source: Johns Hopkins University)

Meanwhile, medical science is devoting huge resources searching for a vaccine and to improve the prognosis for people who catch the virus with over 170 research programmes (47) aimed at producing a vaccine as soon as possible with eleven already in large-scale efficacy trials (Phase 3, the final stage of the development and testing process) and a further nineteen in expanded safety trials (Phase 2) as at 9 October, 2020.

These research programmes are being carried out by scientists using the scientific principles of evolutionary biology. It's probably superfluous to say that there are no reports of so-called 'Creation scientists' or Creation 'science' organisation doing any works whatsoever. Some

potential vaccines are showing promise and Russia has already
approved one ahead of full safety checks.

The optimistic forecast is that we may have one by the end of this year
(2020) but more cautious voices are saying it could be 12-18 months
before we can start mass vaccination programs. At this time there is no
certainty that antibody immunity will be permanent or whether, like the
common cold, immunity will be short-lived.

According to Johns Hopkins Medical Center, COVID-19 symptoms
include:

- Cough
- Fever or chills
- Shortness of breath or difficulty breathing
- Muscle or body aches
- Sore throat
- New loss of taste or smell
- Diarrhea
- Headache
- Fatigue
- Nausea or vomiting
- Congestion or runny nose

In rare cases, Covid-19 can lead to severe respiratory
problems, kidney failure or death (48).

SARS-CoV-2 like all coronaviruses, is a single-stranded RNA (SSR)
virus, closely related to the virus that caused SARS a few years ago.
There is little doubt that it is related to a coronavirus commonly found
in bats. Some virologists believe the number of mutations needed to
produce the SARS-CoV-2 virus is too large to have occurred in a single
species and so at least one intermediate host may have been involved.
The pangolin and feral chickens have been suggested as possible

candidates. To an ID advocate, of course, the only explanation permitted is that the virus was deliberately designed!

The mutations involve the RNA genes that code for the so-called spike proteins that protrude from the viral coat. These act as keys which fit onto proteins on the host cell surface, known to science as ACE2, so enabling the virus to insert its RNA into the host's cells. The ACE2 protein on the cell surface is thought to be involved in regulation of blood pressure and fluid and electrolyte balance, as well as cardiovascular, neurovascular and renal function, and fertility. The shape of the spike proteins on the bat virus are too different to those need for human cells to be caused by a single set of chance mutations, so the argument goes. However, with a single virus being capable of producing trillions of copies in a very short time, this means that even a 100 billion to one chance will arise within population very quickly.

Two scientists working at Northwestern University, Evanston, Illinois, USA discovered just how detailed this re-design work was when they found the virus is helped bind to human cells by a positively-charged area on the virus spike protein matching a negatively charged part of the ACE2 cell-surface protein on a human cell, so helping the virus bind more firmly for maximum infectivity (49). No self-respecting ID advocate would look at that and deny that it was intelligently designed, would they? Clearly the work of a malevolent evil genius!

There is little doubt too that bats are uniquely able to harbour these viruses which are so deadly when they cross the species barrier to humans. They appear to have an immune system which enables the bat to minimise the damage they do to their cells but not to eliminate them. In effect, bats enable viruses to evolve within them and metaphorically experiment with new mutations. As was reported by Stony Brook University News (50), this is now the subject of a research programme by scientists at Stony Brook University, New York (51).

A mutation which seems to have arisen early on in the history of SARS-CoV-2 was one reported recently by researchers from the Theoretical

Biology and Biophysics Department of Los Alamos National
Laboratory, Los Alamos, New Mexico, USA. They showed that a
minor modification of the original D614 form to give the D614G
variant has made the virus more infective so this variant has almost
replaced the original form throughout the world (52).

In ID terms, SARS-CoV-2 has already been redesigned to make it **more**
infections.

According to the American National Institute of Allergy and Infectious
Diseases (NIH):

> There are hundreds of coronaviruses, most of which circulate
> among such animals as pigs, camels, bats and cats. Sometimes
> those viruses jump to humans—called a spillover event—and
> can cause disease. Four of the seven known coronaviruses that
> sicken people cause only mild to moderate disease. Three can
> cause more serious, even fatal, disease. SARS coronavirus
> (SARS-CoV) emerged in November 2002 and caused severe
> acute respiratory syndrome (SARS). That virus disappeared by
> 2004. Middle East respiratory syndrome (MERS) is caused by
> the MERS coronavirus (MERS-CoV). Transmitted from an
> animal reservoir in camels, MERS was identified in September
> 2012 and continues to cause sporadic and localized outbreaks.
> The third novel coronavirus to emerge in this century is called
> SARS-CoV-2. It causes coronavirus disease 2019 (Covid-19),
> which emerged from China in December 2019 and was
> declared a global pandemic by the World Health Organization
> on March 11, 2020 (53).

A virus which ID advocates have to believe was intelligently designed
to cause diseases in several different animals has the potential, with a
small modification, to jump species into humans and cause a serious
respiratory illness. And it gets worse for Creationism. Scientists from
the Francis Crick Institute with colleagues from Aarhus University,
Denmark and the Royal Veterinary College, showed that Sars-CoV-2

induces the cells to produce Interferon Types I and III in response to attack but these, when produced over a prolonged period, actually make it more difficult for the body to repair the damage the virus has done and the longer the infection lasts, the more difficult it is to repair the damage (54). The virus turns the body's immune system against itself!

While this book was in progress, during which there are literally hundreds of research groups trying to get a better understanding of SARS-CoV-2, researchers led by led by Walter J. Lukiw of the Neuroscience Center at Louisiana State University School of Medicine, found that the ACE2 cell-surface protein is widely distributed in the human body including in lung, digestive, renal-excretory, reproductive, eye tissues, and 21 different regions of the brain (55). They suggested that this is why the virus is able to damage such a wide variety of tissues and organs. They also suggested that a possible route for infection could be through unprotected eyes, making eye-protection important in measures to control its transmission.

Another research team, led by Joana Damas of the University of California, Davis Genome Center recently showed that the 25 amino-acid sequence that forms the binding site for the SARS-CoV-2 on the ACE2 protein is also found in many other species (56). This means that these species, including our closest relatives amongst the great apes, and several endangered species, may also be susceptible to the virus and can act as a repository for it even if we eliminate it from the human population. SARS-CoV-2 may well be a highly zoonotic species of virus capable of moving from one species to another with ease. The longer it remains in the wild, the greater the chances of another mutation arising which will make it invisible to any immunity we may have developed. ID Creationists must assume that this was a deliberate design feature of SARS-CoV-2.

Given that Creationist ID advocates believe their putative creator is also omniscient, they can't argue that it could not have foreseen the terrible consequences for human society of its little modification of the SARS-CoV virus to produce SARS-CoV-2.

The only defence they can offer is that SARS-CoV-2 is in some way the consequence of the Christian myth of 'The Fall' when 'sin' entered the world and began a gradual degradation or 'genetic entropy', so abandoning the pretence that intelligent design is real science and not Christian fundamentalism in a grubby lab-coat.

The notion of 'genetic entropy' is nonsensical and designed to make the notion look like science and somehow consistent with the fundamental scientific Laws of Thermodynamics. It is not of course, since there is no known mechanism by which disadvantageous (degraded) genes can accumulate in a species gene-pool and Creationists certainly don't propose one. There is nothing 'degraded' in a mutation that gives an advantage. Of course, the whole basis for claiming 'degradation'. Or 'devolution' as Michael J. Behe calls it, is the unscientific, religious belief that a perfect creator created a perfect creation, so any deviation from that perfection must be 'degradation' in some way, regardless of how much it improves on the initial 'perfection'.

The cause of this 'degradation' is, of course, the unscientific, superstitious notion of 'The Fall' and 'Original Sin'. One can only assume that Michael J. Behe and his Discovery Institute have abandoned the charade of pretending intelligent design and Creationism is real science, not Christian fundamentalism, following their humiliation in Kitzmiller vs Dover District School Board (12).

The entire basis of genetic change, such as that which produced SARS-CoV-2, is random mutation. Whether that mutation proves beneficial or deleterious depends entirely on the environment. In the case of SARS-CoV-2, from the point of view of the virus, the result of that mutation was unarguably beneficial since there must now be countless trillions of copies of it in the world's human population.

But we are playing the part of believers in Creationism's intelligent designer for the purpose of this book, aren't we? So, we have to accept that SARS-CoV-2 and the Covid-19 disease it causes was the intended outcome for the ingenuity of the designer who designed coronaviruses and then modified this one. The suspicion is that it gave bats a unique ability to harbour these viruses where they can mutate and 'experiment', almost like Creationism's intelligent designer has a laboratory for experimenting with new viruses! Or does it just favour bats, so designed them a special immune system which works better than the one it designed for us?

The terrible dilemma Creationists now face is accepting that evolution is the only explanation for SARS-CoV-2 that does not leave their putative designer looking for all the world like a misanthropic, sadistic, pestilential, genocidal, malevolence who is assiduously designing nasty little parasites, the better to make us sick.

But enough about Covid-19. There are plenty of other examples of how any sentient, omnipotent, omniscient designer who reputedly designed all living things cannot avoid the charge of malevolence. The forthcoming chapters contain just a few of the many thousands of such examples.

Creepy Crawlies

Looking at the world through the distorting lens of intelligent design, it must appear that very many otherwise fascinating and exquisite insects were designed for no other purpose than to make life difficult or unpleasant for other things.

I wrote about several of the endo-parasitic wasps and their hymenopteran relatives, ants, in *The Unintelligent Designer,* as examples of unnecessary complexity and prolific waste, both of which are evidence of unintelligent design or lack of design in nature, but what I never looked at were the insects and mites that live as ectoparasites that afflict us and the 'design' features that facilitate their success as parasites. As we have already seen with the black-legged tick, and mosquitoes, they, like many blood-suckers, are also important vectors for several other nasty little parasites. It seems Creationism's intelligent designer has never missed an opportunity to use one parasite to inflict another parasite on us and other creatures.

Scabies.

Scabies has been described as not a disease but an infestation. The cause is a microscopic mite that burrows into the skin where the female lays between 10 and 25 eggs. These hatch into nymphs which move onto the skin and take up residence in hair follicles. The presence of the adults, nymphs, eggs and their faeces is responsible for the intense itching which is often the first sign of infestation. Symptoms can be produced by as few as 10-15 mites.

S. scabiei var. hominis is an obligate, species specific parasite, which can only reproduce successfully on humans. There are other closely related varieties that live on other mammals, including dogs, cats, cattle and the great apes.

Figure 9 Scabies mite, Sarcoptes scabiei

I can attest to the unpleasantness of this little parasite having been infested myself. As a paramedic and well known in my then small home town, I was shopping in a local supermarket when an elderly man was taken ill. One of the staff who knew me, found me and asked me to come and look after him until an ambulance arrived. He was obviously in poor health and suffering from self-neglect. It later transpired that he had an especially infectious form of the infestation – Norwegian or crusted scabies. I recovered completely after treatment.

Because of the intense itching it is difficult to resist scratching, so the skin can become raw and infected with opportunist bacteria.

Note that the variety that infests humans is species-specific. This means, if you believe in a designer, that the mite has been specially designed to live on humans, so the unpleasant symptoms were the intention of the designer. *S. scabiei* appears to have no other purpose than to make more mites and to infest humans to give them intense itching and risk of bacterial skin infections.

Fellow travellers – Lice and Typhus

The lice humans have are, like the scabies mite, *S. scabiei*, species specific, obligate parasites. This means they have no choice but to live on humans. That means also that as humans have evolved over the years, our lice have had to adapt or die.

Humans have three lice:

- *Pediculus humanus capitis*
- *Pediculus humanus humanus*
- *Phthirus pubis*

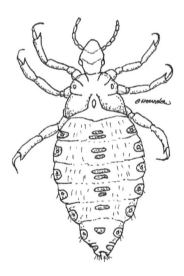

Figure 10 Body louse, Pediculus humanus humanus

The thing is that, while the first two are very closely related, being only subspecies of *Pediculus humanus*, the third, the pubic or crab louse is from a different genus, *Phthirus*. As it happens, both species of chimpanzee also have an obligate parasite louse – *Pediculus schaeffi* –

and gorillas have their own - *Phthirus gorillae*. In other words, we share a genus of lice with our two closest relatives.

What evolutionary biologists argue (57), with genetic evidence to support it, is that as humans and chimpanzees diverged, our dependent lice had to diverge too (58). As humans underwent a loss of body hair, our lice were restricted to our heads. It was only later when we moved out of African into colder northern climates and invented clothing, that a new niche opened up for our lice to diversify again into the head louse and the body louse (59). The two species of Chimpanzee only have the one louse because they never lost their body hair.

At some point, somehow, we managed to acquire the gorilla louse which then had a separate niche in our pubic hair, where it has become a rather nasty little sexually-transmitted species known as 'crabs' or pubic lice. Quite how that inter-species transfer happened can only be guessed at but the genetic evidence again shows the relationship between the gorilla louse, *Phthirus gorillae* and the human pubic louse, *Phthirus pubis.*

Creationists, however, reject that explanation on the curious dogmatic belief that evolution cannot account for the diversification into different species. Of course, they also reject the evidence of common ancestry of humans and the other apes, of which the evidence of the relationship between our lice is supporting evidence. This just leaves them with the idea that all the different lice were specially designed to do what they do – suck blood and live on our bodies, where they are important vectors for diseases such as typhus – something Creationists must believe their putative designer also designed to do what it does – kill people or make them seriously ill. Typhus is the name given to a range of parasitic illnesses caused by organisms injected by blood-sucking arthropods. The organism transmitted by lice is *Rickettsia prowazekii.*

According to the NHS:

Symptoms of typhus include:

- headache
- very high temperature (usually around 40C)
- nausea, vomiting and diarrhoea
- dry cough
- tummy pain
- joint pain
- backache
- a dark spotty rash on your chest that may spread to the rest of your body (apart from your face, palms of your hands and soles of your feet) (60).

Although typhus can be treated with antibiotics this was not always so. It is thought to have been the cause of the first recorded 'plague' – the Plague of Athens in 430 BCE when it killed an estimated 75,000 to 100,000 people and changed the course of the Peloponnesian War. It is also thought to have changed the course of history when it wiped out about half of Napoleon's army with which he invaded Russia.

One instance of a major outbreak of typhus and the first reliable historical report of it, which might be of interest to ID advocates, was during the Siege of Baza in Spain when a Christian army was besieging the Moorish-held town during the conquest of Granada by the Christian north. It killed some 17,000 Christian troops as against the 3,000 lost by enemy action!

I probably don't need to spell out the problems lice and typhus hold for those wishing to present a putative designer as benevolent and anthropophilic.

Bed Bugs.

Bed bugs are blood-sucking insects that are entirely dependent on humans and human habitation, being unknown in the wild and found only in human dwellings. They come is two related species of the Cimex genus, *Cimex lectularis* and *C. hemipterus*. They hide in narrow cracks in walls and floors and emerge at night to feed on people while they sleep.

Figure 11 Bed Bug, Cimex lectularis

The bites can be red and intensely itchy and may produce an allergic reaction. Rarely, a bite can produce anaphylactic shock. The bites can be raised and painful and may become infected by opportunistic bacteria due to itching. Bed bugs have no known benefits.

The evolutionary explanation for bed bugs is that they evolved from bat bugs that inhabited caves occupied by bats. When proto-humans began to use caves as settled habitations, bat bugs evolved to exploit this new resource. Unlike other apes, humans tend to return to the same location and even the same bed to sleep. This makes us vulnerable to parasitic insects, mites, etc, that can hide in the vicinity of our beds. Nesting birds have the same vulnerability as we will see shortly. The ancestors of bed and bat bugs predates both bats and birds by some 50 million years and were around with dinosaurs. As bats and birds evolved, the bugs diversified to parasitize both, but it is from the bat bug that our bed bugs almost certainly evolved.

Genetic evidence shows that bed bugs probably diversified from bat bugs in Africa (the earliest known have been found fossilised in Egypt).

Bed bugs are undergoing something of a resurgence, especially in New York's hotels. On reason for this is that they have evolved resistance to the insecticides used to control them, or, to put that in ID terms, they have been redesigned to be resistant to the chemicals we developed to defend ourselves from them with. Another possible reason is our success at controlling cockroaches, one of the few things that predates on bed bugs.

To an ID advocate, bed bugs are deliberately designed to do what they do and have even been given resistance to commonly used insecticides. Whilst a perfectly rational explanation is offered both for their existence and their insecticide resistance by the science of evolution, an ID advocate would rather attribute it to what can only be described as the malevolent intent of a creative intelligence.

Fleas and Plague.

Unlike many fleas which are species specific, despite its name, the Human flea, *Pulex iritans* has a wide range of host species, including dogs, cats, and black and brown rats.

In addition to the itching that its bites produce, it is also a carrier of several nasty diseases including a tapeworm, *Dipylidium caninum*. It is perhaps most famous for spreading *Yersinia pestis*, the coccobacillus bacterium believed to have been responsible for the Black Death plague which devastated the world in the Middle Ages, peaking in Europe between 1347 and 1351. It was not unusual for the entire population of a village to be wiped out.

Y. pestis produces three forms of plague depending on how it is transmitted. Those caused by flea bites are bubonic and septicaemic plague. The former is characterised by swollen lymph nodes or 'buboes' in the armpits and groin; the latter by the body turning black due to destruction of red blood cells and subcutaneous haemorrhages. Both can be rapidly fatal. An airborne form – pneumonic – is also fatal and can be spread in respiratory aerosols by people infected with either of the other two forms.

The Black Death is estimate to have killed between 75 million and 200 million people in Europe and North Africa, causing profound religious, social and economic changes. It is believed to have originated in Central or East Asia and been carried along the developing Silk Route which had been made possible by the stability imposed by the Mongolian Empire of Genghis Khan a century earlier, reaching Crimea on the Black Sea coast by 1341. It was then spread throughout the Mediterranean basin by fleas and black rats on Genoese trade ships. Human cultural and economic development had created an environment in which rats, fleas and *Y. pestis* thrived.

The Black Death plague is believed to have killed 30%-60% of the world population. Outbreaks continue to occur occasionally but there has not been a serious outbreak since the 19th century. On 17 August 2020, the Los Angeles Times reported that health authorities in California, USA had confirmed the first case of plague in 5 years in South Lake Tahoe (61). The patient was reported as recovering after treatment.

It is difficult to think of any single insect/bacterial combination that has had such a profound effect on humanity. At the time, many religious people interpreted the Black Death as God's punishment for some assumed transgression – not unlike modern Creationists who can always point the finger at something they don't like, such as same-sex marriage, legalised abortion or votes for women and declare that God shares their displeasure.

The fact that *Y. pestis* killed the pious and impious alike and in equal measure, simply indicated that either God's displeasure was with humanity overall, that the pious were doing it wrong, or that God's mind is unknowable and he 'works in mysterious ways'. The perception that the Catholic Church had become corrupt and debauched and so 'caused' the plague, may have facilitated the acceptance of radical change such as that which led to the Protestant Reformation.

In 2015, a team of Danish researchers showed that *Y. pestis* had been common in Eurasia in the Bronze Age, but was to undergo a modification before 951 BCE to make it transmissible by fleas. The Bronze Age form had been incapable of producing Bubonic Plague (62).

The ID interpretation of this modification must be that it was the work of an intelligence for the purpose of making *Y. pestis* capable of causing Bubonic Plague and to be transmissible by fleas, and so causing the massive devastation that ensued. Again, an interpretation that leaves the putative designer looking malevolent, pestilential and mendacious, simply to avoid the scientific, evolutionary explanation – that a chance mutation gave the organism a significantly greater success in the prevailing environment.

Sand Fleas or Jiggers

Before I leave the fleas, I will mention one more – a nasty little, debilitating pest common in tropical parts of the world – the chigoe, chigoe flea or jigger, *Tunga penetrans* – a parasitic insect native of South and Central America that has been inadvertently introduced to Sub-Saharan Africa. An infestation of T. penetrans is known as tungiasis. It can parasitize a large number of mammalian hosts. T. penetrans is just one of 13 species of the *Tungidae* genus, all of which are parasitic on mammals.

T. pentrans normally live in the top 2-5cm of sand, hence their other name, the sand flea or sand chigoe. When a virgin female finds herself on a naked human foot, she borrows into the host skin, leaving only the tip of her abdomen exposed, ready to mate with any passing male. Here she lives off the host's blood. Males are also blood-sucking parasites and remain mobile after feeding. After mating, the female's abdomen can swell up to 1cm wide. After laying her eggs onto the host's skin, the female dies, leaving her dead body in the skin where it can be the source of opportunistic bacterial infections. The swelling of the female body from about 1mm to 1cm can cause intense itching and pain due to pressure on nerves and skin blood vessels. Infestation can lead to loss of toenails and toe deformation. Left untreated, infestation can also lead to septicaemia, tetanus or gangrene.

Chiggers.

Chiggers are microscopic mites, relatives of ticks, that live in vegetation and attach themselves to passing hosts including humans. The commonest one in North America is *Trombicula alfreddugesi* while that in Europe, including the UK, is *Trombicula autumnalis* or 'harvest mite'. The adult form of the mite is harmless but the larval stage is parasitic, living on the skin cells of its host. Unlike blood-sucking arthropods, the *Trombiculidae* inject digestive enzymes into the skin of their hosts and eat the resulting cell debris. This causes trombiculosis - intense itching and dermatitis which, in humans, develops 24-48 hours after infestation.

Making a Better Mosquito.

Parts of this next article are taken from a blog post I wrote when this news was first announced (63).

Mosquitoes like *Aedes aegypti,* which are often Creationism's Intelligent designer's blood-sucking parasite of choice when it has another nasty little parasite like Zikavirus or *Plasmodium falciparum* to deliver into our blood stream and make us sick and die. So, as we've come to expect of the malevolent intent of this putative designer, it takes a belt and braces approach in its designs leaving little to chance in its fanatical zeal.

For example, as researchers from the Department of Biological Sciences & Biomolecular Sciences Institute, Florida International University, Miami, FL, USA have shown, the antennae of female *Ae. aegypti* mosquitoes can 'smell' characteristic signature chemicals in human body odour, so they are more able to find their targets, like an Exocet missile (64).

The *Ae. aegypti* mosquito originated in Africa but is now found in tropical, subtropical and temperate regions throughout the world. It can carry yellow fever, dengue, chikungunya, Mayaro and is almost single-handedly responsible for the spread of the Zikavirus which causes microcephaly in children if their mothers become infected during pregnancy.

The Florida International University team blocked a gene responsible for the ir8a co-receptor on the antennae of females and found that they were 50% less able to detect their hosts to obtain a blood meal.

The evolutionary explanation for this ability of a predator to find its prey is self-evident. Anything that improves that ability will give an advantage which will be passed on to the next generation in greater numbers than those without it.

For Creationists, however, we have again the problem of explaining why an intelligent designer would go to these lengths to enable its designed delivery system to more effectively find humans and other warm-blooded animals and deliver their cargo of life-threatening parasites into their blood. Did this intelligent designer give *Ae. aegypti* this ability so babies could be born with microcephaly, or so people could die of yellow fever or dengue. The inescapable conclusion for an ID advocate is that this is exactly what it did, given that they believe this creator knows precisely what its designs will do and designs them to do precisely what they do. None of this could happen against the will of an intelligent designer who, even if it made a silly mistake, should be more than capable of undoing it.

So far, I've tended to concentrate on parasitic arthropods that are parasitic on humans. This of course ignores the very may examples of parasites on other creatures, many of which are examples of 'design' that can only be regarded as mendacious, if the work of a sentient designer. I will look now at some of the parasite-host relationships that are impossible to see as the work of a benevolent intelligence.

Lycaenid Caterpillars and ants.

Caterpillars of the *Lycaenid* butterfly, *Narathura japonica* secrete a nutritious substance from a dorsal nectary organ which is eaten by worker ants of the *Pristomyrmex punctatus* species, which then protect the butterfly larvae from the attention of parasitic wasps. This might, on the face of it, look like a simple symbiotic relationship where a reward is given for service. However, a closer examination shows that the caterpillar is actually controlling the behaviour of the ant by feeding it on dopamine which makes the ant more aggressive and more inclined to remain close to the caterpillar rather than returning to the nest.

This was discovered by a team from the Faculty of Agriculture, University of the Ryukyus, Nishihara, Okinawa, Japan and Museum of Comparative Zoology, Harvard University, Cambridge, MA, USA who found that the brains of these ants contained much higher concentrations of dopamine than normal, showing that the caterpillars were manipulating the ants and modifying their behaviour. (65)

From an ID view, what we have in this example is a parasitic wasp that is designed to lay its eggs in the bodies of living caterpillars, then a 'solution' to that deliberately created 'problem' that involves the chemical manipulation of a third species so it attacks the parasitic wasp. The victim in all this is the ant that was doing what it was designed to do and foraging for food to take back to the colony.

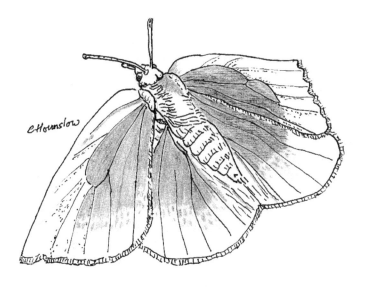

Figure 12 Lycaenid butterfly, Narathura japonica

Killer Bumblebees.

The 29 species of parasitic bumblebees, of the *Psithyrus* subgenus of the *Bombus* genus, have given up making their own nests and living socially with a worker caste. Instead, they invade the nest of a social bee such as the buff-tailed bumblebee, *Bombus terestris* and kill or subdue the queen and lay their eggs in the cells prepared by the host workers. She then 'enslaves' the workers, using pheromones and by attacking them, to feed her and her developing young. Her eggs develop into both male and female adults.

When her young reach adulthood, they leave the nest and disperse, while she seeks out other nests to invade, as there is now no queen to replace the aging workers (66) (67).

Zombie Caterpillars and Parasitic Wasps.

The caterpillars of the Geometrid moth, *Thyrinteina leucocerae*, are heavily parasitized by endoparasitoid wasps of the *Glyptapanteles* genus which sting the caterpillars and then lay up to eighty eggs in their living body. The caterpillars recover from the sting and continue to feed as normal. The larvae of the wasp have meanwhile been eating the insides of the caterpillar, leaving only the vital organs.

When the wasp larvae are fully grown, they exit through holes in the caterpillar's sides and pupate nearby. However, one or two larvae remain in the caterpillar where they manipulate its behaviour in an unknown way so that the caterpillar spins a protective covering over the pupae and violently swings its head from side to side to defend the wasp pupae from predators. The unfortunate caterpillar victim actually defends the pupae of the wasp that has effectively ended its life. This behaviour has never been observed in an un-parasitized caterpillar (68).

Let's run through what is going on here from an ID view. A designer has designed a moth and then designed a wasp to treat the larval stage of that moth as a resource to be exploited to produce more wasps, not more moths. Then it designed predators to eat the pupae of the wasp, so, to solve the problem **that** caused, the wasp was redesigned to manipulate the caterpillar victim to prevent that predator from doing what it was designed to do. And all of this was by an intelligent designer who knew exactly what its designs would do.

ID advocates regard that as a much more reasonable explanation for those facts than that offered by science with the Theory of Evolution by Natural Selection.

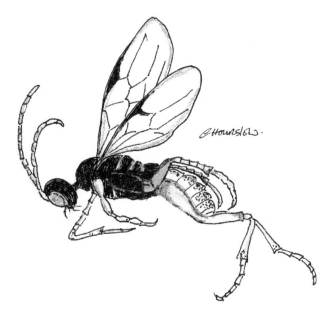

Figure 13 Glyptapanteles compressiventis

The Braconids are a vast group of the hymenoptera comprising as many as 50,000 species. Very many of these are parasitic on other insects, usually, but not always on the larval stage or caterpillar.

Braconids and Biological Warfare.

In 2015 a team or researchers from Universitat de València, Burjassot, Spain and Université François Rabelais, Tours, France found that several lepidopterans (moths and butterflies) had sections of DNA from Braconid wasps, even though the two orders diverged some 300 million years ago. These were obtained by horizontal gene transfer (69).

In order to supress the caterpillar's immune system so it doesn't attack the wasp's larva growing inside it, the wasps also transfer a large virus known as a bracovirus, which integrates with the caterpillar DNA and produces substances that supress their immune response. These DNA sequences are actually symbiotic remnants of retro-viruses that probably entered the ancestors of the wasps early in their evolutionary history and have since become endogenous retroviruses that have then been co-opted by the wasps. The virus particles are produced in the wasp's ovaries. In addition to the virus, the wasp sometimes also transfers some of its own genes that give the caterpillar some defence against a common baculovirus.

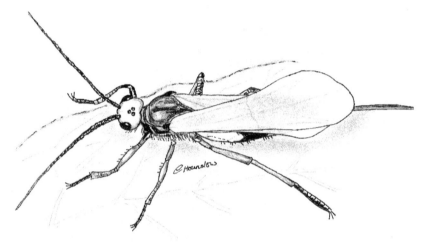

Figure 14 Yellow-flanked braconid wasp, Chaoilta hollowayi

The fact that these wasp genes and bracovirus DNA are now present in lepidopterans suggests that occasionally, infected caterpillars have survived the wasp parasites and produced adults, complete with the bracovirus and wasp DNA in their germline cells. Because these give the descendants protection against the baculovirus, these have increased in the species gene pool. In effect, braconid wasps have genetically engineered their hosts, using a virus to insert genes into them in much the same way that human genetic engineers do.

An evolutionary explanation for this phenomenon is easy to understand as the result of evolution by natural selection but what is not easy to understand is why an intelligent designer would use such a complex way to make moths and butterflies resistant to the baculovirus it designed to attack them with or why it would adapt a virus it had inserted hundreds of millions of years ago into the ancestral wasp's genome to suppress the immune responses it gave to caterpillars to defend themselves with, so the grubs of a wasp can eat it alive.

The Parasite That Controls Spiders

Reclinervellus nielseni is an ectoparasitic wasp that takes control of an orb-web spider and makes it build it a safe web in which to pupate. A female *R. neilseni*, a wasp belonging to the *Polysphincta* genus, seeks out orb-web spiders of the *Cyclosa* genus and stings them to make them harmless. Then it either injects an egg into their abdomen or, in the case of *Cyclosa argenteoalba,* found in Japan and common around shrines, since it prefers to build its webs in man-made structures, it glues an egg on the spider's abdomen.

The resulting grub then rides on the back of its spider host, feeding off its body fluids through small holes it makes in its body wall, until it is large enough to pupate.

These spiders normally make two types of web: a typical sticky orb-web for catching flying insects and a shelter 'rest web' for when it needs to rest and moult its skin. This web has no sticky strands as it is not meant to catch insects but to protect the spider. For a short while after they moult a spider is soft and vulnerable. This rest web is decorated with spiral patterns to make then visible, presumably so they are not accidentally destroyed by birds and larger flying insects. Some of the decorations reflect ultra-violate light, to improve their visibility.

When the parasite is ready to pupate, it induces the spider to build an especially robust version of this 'rest' web with more decoration and especially strong radial strands. Unlike with a moulting spider, which is only needed for a few days, the wasp larva needs the web to last for the seven to ten days it takes to metamorphose into an adult wasp!

As soon as the special web has been built and the spider taken up residence in the centre of it, the larva kills the spider by sucking it dry. It then casts its dead body away and spins a cocoon in the centre of the web.

How the wasp grub induces the spider to build this special web is not known with any certainty, but it is assumed to injects the spider with a version of a hormone the spider will normally produce prior to moulting, which stimulates building the rest web. The fact that parasitized spiders build stronger and more durable rest webs for the parasite that is destroying them, so cooperates in its own destruction, was show by a team of researchers, led by Keizo Takasuka from Kobe University, Hyogo, Japan (70) who gathered spiders from a nearby shrine and observed them in a laboratory setting.

It's almost as though whatever designed the spider, saw the mechanism it had designed for it so it could moult safely, as a golden opportunity when it wanted to keep its parasitic wasp larvae safe while they made more parasitic wasps. Imagine the character of a designer who would do such a thing to its creation! But then, host manipulation is commonplace in the parasite world. ID advocates are, of course, obliged by their dogma to assume this was intelligently designed and fully planned by their putative designer who seems to have a fascination and predilection for parasites and parasitic wasps in particular.

We have seen several instances of hymenopterans being parasitic on butterflies and moths, normally in their larval stage. I will look now at a situation where the tables are turned and a butterfly is parasitic on a hymenopteran, in this case the red ant.

Large Blue Butterfly and Red Ants.

The large blue butterfly (*Phengaris arion*) is a rather beautiful blue butterfly of the *Lycaenidae* family which lives (or rather lived – it is was extinct in UK, but has been the subject of reintroduction efforts) on southern chalk hills.

Figure 15 Large Blue, Phengaris arion

Its caterpillars feed on wild thyme and marjoram until they are large enough, then they trick red ants of the *Myrmica sabuleti* species to take them into their nests. It is not clear how they do this but chemical and other mimicry is probably involved. Once in the nest they adopt one of two strategies – they either beg for food like the ant larvae (the 'cuckoo strategy) or they become predators on the larvae.

When ready, the larvae pupate in the ant nest where they are protected and cleaned by the ants as though they are ant pupa. The existence of the *P. arion* larva in the ant nest is a precarious one and many are attacked and eaten. To avoid this, the larvae use chemical and auditory mimicry to enhance their status within the colony by mimicking a queen, including the sounds she makes. However, this strategy has attendant risks since an established queen herself has a strategy for preventing any of the potential queens developing to be in a position to challenge her and a healthy, egg-laying queen will not tolerate a potential rival lightly.

Queens lay two batches of eggs; ones that will hatch into sterile workers and ones that, if allowed to, will hatch into virgin queens and males. If the queen is strong and healthy, she produces a substance which induces the workers to starve, neglect and bite potential rivals, so they fail to develop and hatch as workers. *P. arion* larva must therefore reach a balance between rivaling the queen and being recognized as an alien and potential food.

The Tongue-Eating Louse.

Some parasites are the stuff of nightmares, yet, if ID advocates are to be believed, their putative intelligent designer spends its time designing these creatures for no discernible purpose save to make more copies of themselves. One such is the so-called tongue-eating louse, *Cymothoa exigua*, which is not a louse at all, though it is an arthropod. It is an isopod of the order *Cymothoidae*.

The female of this parasite enters a fish mouth via its gills and attaches itself to the base of the fish's tongue. It then uses its front claws to sever the blood vessels supplying the tongue, so the tongue atrophies and falls off. It then attaches itself to the muscles in the stub of the tongue and effectively replaces it. A male isopod attaches itself to a gill arch immediately behind the female. This is the only know case of a parasite effectively replacing an organ in its host (71).

Once attached, some isopoda live on fish blood, while many others live on mucous.

C. exigua is not the only member of this genus to pull this trick. *C. borbonica* and the related *Ceratothoa imbricate* also replace the tongue of their host fish. C. exiguais is currently known to parasitize eight different species of fish but there are signs that it may be extending its range both into new species (72) and geographically. A red snapper caught in UK waters was found to have an isopod in its mouth, although

70

it could have been carried there by the fish rather than the fish being parasitized in UK waters.

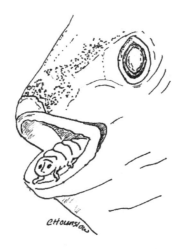

Figure 16 Cymothoa exigua in mouth of a fish

The Crab-Castrating Barnacles.

The genus of barnacles known as *Sacculina* bear little resemblance to the familiar barnacles seen encrusting wooden piers and the hulls of ships and are only recognised as members of this group of arthropods by their larvae. The commonest is *Sacculina carcini* but there are over 100 species.

They are parasitic on crabs, primarily the green crabs of the Eastern Atlantic. Up to fifty percent of crabs can be infected.

When a female *Sacculina* larva finds herself on the body of a crab, she finds a joint in the shell and injects the soft part of her body into the body of the crab where she grows, emerging as a sac-like structure at the rear of the crab's thorax under the tail, where eggs would normally be brooded. Extending from this into the body of the crab are root-like tendrils that wrap themselves around the crab's organs and brain, drawing nutrients from them and controlling its behaviour.

Then, *Sacculina* destroys the crab's gonads. If the crab is a male, it femininizes it, making it produce hormones that alter its body shape to resemble and behave like a female. This also induces the nurturing behaviour of a female crab who would normally protect her eggs beneath her tail until they are ready to be dispersed.

Figure 17 Sacculina carcini under the tail of a green crab

Meanwhile, a male *Sacculina* larva seeks out a virgin *Sacculina* on the underside of crabs and, when he finds one, he injects cells into a special pocket where they develop into spermatozoa which fertilise her eggs.

An uninfected female crab will shed her eggs into the water from a rock by shooting them out in batches while stirring the water with her claws to aid dispersal. An infected one will go through the same performance, only in place of eggs, she is dispersing *Sacculina* larvae, to repeat the cycle of infection over again (73).

ID advocates would have us believe this parasitic barnacle was specially designed to turn green crabs into automatons for producing copies of *Sacculina* instead of copies of green crabs.

Louse Flies, Swallows, Martins and Swifts.

Just one more unpleasant little arthropod – in this case a highly modified flightless fly - before we leave the world of arthropod horrors. This one exploits the tendency of some birds like swallows (*Hirundo rustica*), house martins (*Delichon urbicum*) and swifts (*Apus apus*) to return to the same nest sites year after year, in much the same way that bed bugs exploited our tendency to return to the same place and bed to sleep.

As a teenager, I was amazed one day when a house martin fell out of the sky, bounced off our house roof and landed next to me. It was quite dead, although whether it died from the fall or died in flight it was impossible to tell. Picking it up, I was immediately struck by a number of largish, wingless insects that were scuttling in and out of its feathers and onto my hands where they clung tenaciously with what looked like grappling hook feet.

These were flightless, blood-sucking parasitic louse flies (*Crataerina hirundinis*) and could have contributed to the bird's death. They are highly modified flies of the *Hippoboscidae* family, all of which are obligate, blood-sucking parasites, mostly on mammals and birds. *C. hirundinis* lives on house martins (mostly) as well as sand martins and swallows. A closely-related species, *C. pallida*, parasitizes swifts. An adult *C. hirundinis* takes about 4.8 mg (females) and 3.5 mg (males) of blood a day so a large infestation can take a heavy toll on the bird. Eggs are laid in the nest of birds where they pupate and remain over winter, to emerge when the birds return in the spring. Their entire purpose seems to be to make more louse flies.

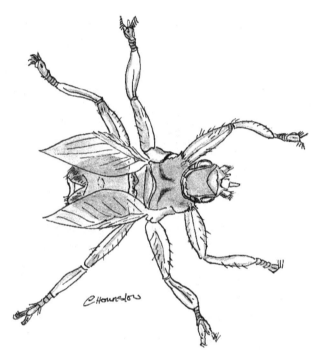

Figure 18 Louse fly, Crataerina palida

Epomis Beetles and Frogs

Epomis are a genus of about thirty species of ground beetles, the larvae of which are obligate, role reversal parasites on amphibians. This behaviour was described in a 2011 open access paper by Gill Wizen and Avital Gasith, of Tel Aviv University Zoology Department, who observed the behaviour of two species, *E. circumscriptus* and *E. dejeani*. Their online paper (74) includes links to several videos of *Epomis* larvae attacking frogs.

Their mode of attack is to pretend to be the food frogs normally prey on, enticing the frog to attack by waving their antennae and mouthparts, then reversing the attack and clinging on to the frog with powerful double-hooked jaws while they eat it alive. The enticing movements intensify as the frog approaches. The *Epomis* larva eventually kills the much larger frog.

If the unfortunate frog succeeds in swallowing the beetle larva, the larva attacks the frog from the inside, causing it to regurgitate the still-living larva, which survives the experience unharmed since frogs have no teeth and swallow their prey whole. Once ejected, the larva immediately grabs the frog, usually near the mouth, and proceeds to eat it. A larva can consume an entire frog before pupating and turning into an adult beetle.

Until their first moult, the *Epomis* larvae are obligate parasites on frogs, detecting their victim by their scent and finding them by moving up the scent concentration gradient. They then pierce the frog's skin and suck the blood through a straw-like mouth-part. After the first moult, they adopt the predatory flesh-eating life-style. At each moult they drop off their host, then find a new victim to attach themselves to.

As adults, *Epomis* beetles continue their predatory life-style, seeking out and attacking frogs, usually from behind, normally by clinging on to the frog's rear legs. They cling tenaciously to the frog with powerful jaws, so the frog is unable to dislodge them, while they eat into the flesh, to find and sever the nerves controlling the leg muscles, so paralyzing the frog. Often, several beetles join in the attack, eating the frog alive and opening up the abdominal cavity to get to the internal organs.

Evolutionary biologists explain this life-style as evolving from an aggressive defence to predation by frogs. ID advocates are obliged by Creationist dogma to see it as the deliberate design of an intelligent designer who presumably saw the frogs it had designed, as potential food for its *Epomis* beetles, but had no regard for the suffering this might cause to the frogs. *Epomis* beetles are just some of the possibly 350,000 known species of beetle (75) – a figure which is constantly rising due to the discovery of more and more different species. The actual figure may well exceed half a million. Creationism's intelligent designer obviously has a fondness for beetles and designed just about every possible variation on a basic theme. The *Epomis* genus is probably one of its more malevolent designs.

That then is just a few of the very many nasty little parasitic arthropods and how they have been supposedly designed then modified to make them even more effective at making us and other creatures sick. The next chapter will look at some worms that do what, to a Creationist view of the universe, must seem like malevolence for the sake of it, since there appears to be no reason for either the organisms or the suffering they cause.

Worms That Make You Squirm

Worms, both round (helminths) and flat (platyhelminths) are a rich source of examples of what, if we accept the intelligent design view, can only be described as malevolent intent on the part of any assumed creator. One estimate has put the number of parasitic worms at 75,000-300,000 species with some 300 parasitic on humans (76).

I wrote about several of them in "*The Unintelligent Designer: Refuting the Intelligent Design Hoax*" because so many of them are examples of prolific waste and of unnecessary complexity. Here I will explain how they are also examples of the lengths a putative designer would have to have gone to, for no other discernible reason than to make us and other creatures sick.

In 2016, Rick M. Maizels, PhD. of the University of Glasgow and Henry J. McSorley, PhD of the University of Edinburgh, showed that people suffering from infestations of worms have reduced immune responses, making them susceptible to opportunist bacterial infections, such as the organism that cases Tuberculosis, *Mycobacterium tuberculosis,* and reduced antibody production in response to vaccines (77). The worms secrete substances that regulate the immune systems of their hosts.

Filaria or Elephantiasis.

Filaria , or elephantiasis is caused by parasitic worms, *Wuchereria bancrofti* which settle in the lymphatic nodes, causing the part normally drained by the affected lymphatics to be hugely oedematous, hence its common name. It affects over twenty million people in Central Africa, the Nile Delta and South and Central America. Inside a human host, male and female worms are tightly coiled together. The female can produce thousands of juvenile 'microfilariae' which live in the circulation, and move between the deeper circulation by day and the peripheral circulation by night – probably taking cues from the host. This makes them available to the secondary host – a night-feeding mosquito – *Anopheles*, *Culex*, *Aedes*, *Mansonia* or *Ochlerotatus*. In the mosquito, the microfilariae mature into larvae which are then injected back into a human host when the mosquito next takes a blood meal. In the human host, the larvae migrate to the lymphatics, usually of the lower limbs and (in men) the scrotum, where they mature into adult worms and mate.

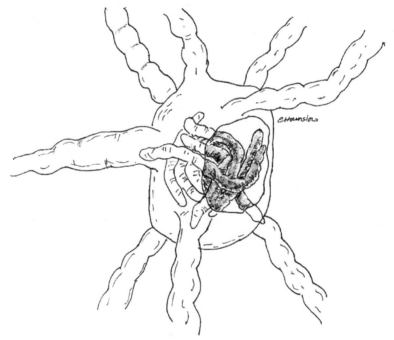

Figure 19 Filaria in lymph node

In Haiti, it has been shown that children of infected mothers are 2-3 times more likely to become infected because they also have lowered immune responses (78). The parasites have also suppressed their immunity in the uterus!

People who are infected with filaria, as with other worms, are also more susceptible to Tuberculosis, caused by *Mycobacterium tuberculosis*, as was shown by researchers from New Jersey Medical School (79).

I doubt that I now need to point out how these facts reflect on the character of any designer who could come up with this system – the same putative designer who allegedly designed the human immune system, *M. tuberculosis*, Mosquitoes and *W. bancrofti*.

The Nematode and Suicidal Crickets.

The nematode worm, *Paragordius tricuspidatus*, in its larval stage is microscopic but grows into a large worm, 10–15 cm long, inside its host, the wood cricket, *Nemobius sylvestris*.

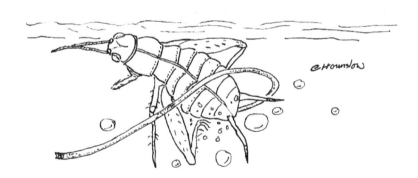

Figure 20 Paragordius tricuspidatus emerging from drowned cricket

Eggs are accidentally ingested by wood crickets when they take a drink from the margins of ponds which are the preferred habitat for the nematode. Inside the host, *P. tricuspidatus* consumes the cricket's inside, leaving only the vital organs, until it fills the entire body cavity.

When the nematode is ready to emerge from the body of the cricket, it induces a bizarre behaviour in which the cricket seeks out water and leaps into it, where it might well drown or end up as a meal for a predator such as a fish or a frog, effectively committing suicide. But dead or alive the nematode exits the cricket's body. If the cricket is eaten, the mature nematode passes unharmed through the digestive system of the predator. Either way the nematode ends up in the water, where it can mate and lay more eggs to repeat the cycle, all to make more *P. tricuspidatus*.

P. tricuspidatus is not the only nematode to pull this trick; a related nematode, *Spinochordodes tellinii*, does the same thing to grasshoppers. Researchers have shown that this nematode produces proteins that influence neurotransmitter activity in its host's brain. These proteins are similar to other insect proteins suggesting chemical mimicry is involved (80).

The Nematode and the Berry-Mimicking Ant.

A recently discovered South American nematode (*Myrmeconema neotropicum*) has an ingenious way to reproduce. It all begins when a foraging ant of the species *Cephalotes atratus,* find a *M. neotropticum* egg on the forest floor, and takes it back to the colony, where it feeds it to an ant larva.

Inside the juvenile ant's gut, the egg hatches and the worm migrates to the gaster where it begins to reproduce. The males die soon after mating but the females retain the eggs within them where they begin to develop. When the ant matures into an adult, the gaster becomes translucent, allowing the red colour of the embryonic nematodes developing within the eggs to show through. The host ant is also induced to go out foraging rather than stay in the nest to do housekeeping duties, as young ants normally do for the first few days.

When the ant ventures outside it has a bright-red abdomen that resembles many of the red berries of the forest canopy. The ant is induced to hold its abdomen up and to become sluggish, so it is mistaken for a red berry by frugivorous birds. Inside the bird's gut the eggs are released and defecated out to repeat the cycle (81).

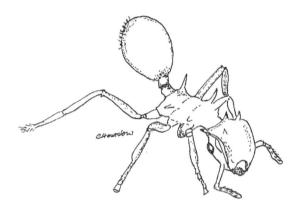

Figure 21 Infected ant, Cephalotes atratus

The Green-Banded Broodsac.

The charming little parasitic flatworm, *Leucochloridium paradoxum,* is parasitic on birds where it lives and reproduces in their digestive tract. Like many parasites that live in the digestive or excretory systems of their host, this means the eggs get dumped outside the host's body and need to get back in to live and reproduce all over again.

This particular one uses snails as the intermediate host. Snails will eat bird droppings, including any *L. paradoxum* eggs. Once inside the snail's digestive tract the eggs hatch into a larval miricidium and then into a long, tubular sporocyte containing hundreds of cercariae (immature worms). The sporocytes migrates to the snail's tentacles, for some reason seeming to prefer the left tentacle, where they transform the tentacle into a colourful, pulsating red and green, banded structure that looks like a caterpillar or insect grub.

Its presence in the tentacle blinds the snail and so makes it insensitive to light, so it is more likely to stay in the open in bright sunlight when uninfected snails would seek shelter. The cercariae however have a pair of eye-spots and **are** light-sensitive. In bright light they become **more** active and pulsating; their activity being directly proportional to the brightness of the light.

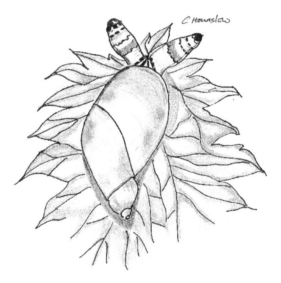

Figure 22 Snail infected with Leucochloridium paradoxum

A 2013 study in Poland by two scientists from Wrocław University showed that 53% of infected snails stayed in the open, occupied higher locations and stayed in better lit places where they were more likely to attract the attention of birds (82). Not only do they blind the snail host, they make it present itself to be eaten, and increase the chances of that happening by turning their tentacles into colourful structures that look like a tasty meal.

This makes them more likely to be spotted by a bird which eats the tentacle, so taking the cercariae into their digestive system to repeat the cycle, all to make more *L. paradoxum,* apparently.

If you can imagine some evil genius trying to think of ever-more ingenious ways to produce these sorts of exquisitely nasty things, then you pretty much have what must be an ID advocate's view of their putative intelligent designer, if they ever have enough knowledge of the reality of biological life to appreciate it.

The Fish-Controlling Tapeworm.

Schistocephalus solidus is a parasite on fish, fish-eating birds and rodents. It has a lifecycle involving two intermediate hosts. Like the previous example, *L. paradoxum, S. solidus* lives in the digestive tract of its definitive host – a fish-eating bird – so its eggs get deposited outside its hosts body and it needs a strategy for getting back in again. When the eggs get deposited in water, they hatch into the first stage – the coracidium. These get themselves ingested by a tiny crustacean, a copepod that goes by the name, *Macrocyclops albidus.* Inside this host they develop into the second larval stage until they are ready to move into the second intermediate host, the three-spined stickleback, *Gasterosteus aculeatus.*

To get themselves eaten by a stickleback when they are infective, the tapeworm manipulates the behaviour of the copepod so it becomes more active. Until that time, however, the copepod is manipulated to avoid becoming fish-food too soon. Experiments have shown that when more than one tapeworm is present in the copepod, they will cooperate if they are both at the same stage of development. If, however, one is infective, it will sabotage the defensive strategy of the non-infective tapeworms and make the copepod active (83).

In some populations, up to 93% of sticklebacks might be infected with *S. solidus.*

Once in the stickleback the third larval stage, the plerocercoid, develops in the fish's abdomen until they are ready to pass into a fish-eating bird and so complete the cycle. *S. solidus* is believed to manipulate its fish host by modifying its predator-avoidance strategies. Infected fish with plerocercoids ready to infect birds were slower to swim for cover when threatened and quicker to emerge from cover than non-infected fish. They also displayed a lower frequency of 'staggered dash' swimming and a higher frequency of 'slow swimming', so making them more likely to be eaten (84).

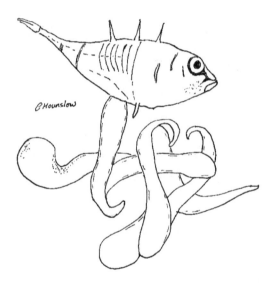

Figure 23 Schistocephalus solidus emerging from the body of a three-spined stickleback, Gasterosteus aculeatus

European Eels and the Swimbladder worm.

While on the subject of fish parasites, another nematode is displaying evidence of considerable 'redesign' to make it more successful at making life difficult for the European eel, *Anguilla anguilla*.

The parasitic nematode, *Anguillicoloides crassus,* in its native range parasitizes the Japanese or Pacific eel, *A. japonica.* Its lifecycle begins in the swimbladder of its definitive host when a mated female lays thousands of eggs which hatch into larvae which pass into the eel's digestive tract and are expelled into water. To get back into the swimbladder of an eel, the larvae are eaten by an intermediate host such as small copepods or other crustaceans, or maybe a fish. Here they mature into their infective form. If the intermediate host is eaten by an eel, the larvae migrate from the digestive tract into the swimbladder, where they mate and lay eggs to repeat the cycle.

In the 1980s, however they were accidentally introduced into Europe, probably from Taiwan. Here they have become serious parasites on the European eel. This process seems to have been made easier by three changes:

Figure 24 Anguillicoloides crassus removed from swimmbladder of European eel, Anguilla anguilla

1. The nematode may be eaten by other species in which it remains immature – the so-called paratenic host. These can include other fish, snails, insects and amphibians, creating a wider range of repositories for the nematode to survive in. Infected copepods tend to be sluggish and inhabit 'epibenthic' zones, i.e. just above the bottom of the body of water, where they are likely to be eaten by benthic (bottom feeder) fish.

2. The nematodes have the ability to infect several different species of copepod, including estuarine species such as *Eurytemora affinis*. This means transmission to the eel can take place in a wide range of aquatic environments.

3. Unlike in the Pacific eel, the larvae stay for much longer in the swimbladder and develop into a fourth stage larva before entering the digestive system and being expelled.

Infected eels may be so heavily infected that their swimbladder cannot function and they are unable to migrate to their spawning grounds in the Sargasso Sea. Consequently, the European eel is now critically endangered, its numbers being reduced to by between 90% and 98% of its 1970s population.

What we are to believe, according to ID advocates, is that the changes that have facilitated this depredation on European eels must be attributed to the deliberate design of a sentient designer who redesigned *A. crassus* for the purpose, because according to the dogma and to force-fit the facts into the preferred religious myth, this could not be the result of evolution by natural selection as a species evolved to exploit a new niche.

To end this chapter, few worms, apart from those which cause blindness in children, which I have written about in *The Unintelligent Designer,* deserve their own section, than the next worm.

The Guinea Worm.

A person becomes infected with the Guinea worm, *Dracunculus medinensis,* when they drink water containing water fleas that are carrying the larval form of the worm. The water fleas are killed in the stomach, releasing the larvae which penetrate the stomach wall and enter the peritoneum. Here, male and female worms pair up and mate. The male then dies but the female migrates to the subcutaneous tissues and grows to about 2 mm wide and up to 60-100 cm long. After about a year she moves around in the subcutaneous tissues, often causing intense pain and eventually migrating to the lower leg or foot where she forms a painful, burning blister which can be relieved by immersion in water. When the foot is in water, the female emerges from the blister and begins to shed eggs which hatch and infect water fleas to begin the cycle again.

Mature worms can be removed from the body by slowly rolling them round a matchstick over the course of about two weeks. The resulting wound may ulcerate and may become infected with opportunist bacteria such as tetanus. Secondary infections have necessitated limb amputations (85).

Its spread can be prevented by drinking clean water or water that has been filtered through cloth to remove any water fleas.

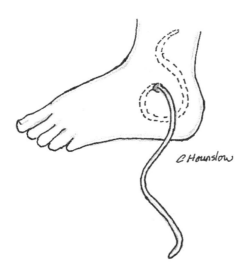

Figure 25 Guinea worm emerging from a foot

Since 1983 the number of cases worldwide has fallen from 3.5 million to just 54 reported cases in just 4 countries, by 2019. There is hope that it may become the first parasitic worm to have been exterminated. However, it has recently been discovered that *D. medinensis,* can also infect dogs, cats, frogs, catfish and baboons, making eradication campaigns more difficult (86).

Could *D. medinensis* be one of the rare instances, along with smallpox and rinderpest, of human medical science defeating what Creationists insist is an omniscient, omnipotent, intelligent designer or will the ingenious malevolent designer come up with a solution to this threat? That remains to be seen.

Making a Better Liver Fluke.

As I wrote in *"The Unintelligent Designer"*:

> *The liver fluke is a parasite on herbivores such as cattle and sheep and humans. It has a complex life cycle which begins when a female migrates from the liver of the host to the biliary duct where she can lay up to 25,000 eggs per day. The eggs are discharged into the intestine of the host and are discharged in faeces. If they land in water, they hatch into a larval form known as a miracidia. The miracidia seek out and infect a water snail.*
>
> *In the snail it goes through three stages, turning from a sporocyst to a redia and finally a large–tailed cercaria which exits the snail and finds aquatic vegetation in which it forms a hard-cased cyst, or metacercaria. If the vegetation is eaten by cattle, sheep or humans, the metacercaria, having been protected against the stomach acids by its hard case, emerges from the cyst in the duodenum and burrows through the intestinal wall into the peritoneal cavity. From there, it*

migrates to the liver and spends time eating liver cells until ready to breed and repeat the cycle (1 p. 35).

We can see then that any putative designer has already gone to considerable lengths to ensure his design, the endoparasitic platyhelminth, *Fasciola hepatica*, ends up in the right place to make its intended victims sick, but as scientists working at the University of Córdoba, Spain, have discovered, it left nothing to chance in its fanatical zeal.

What they discovered is that as soon as the fluke is inside its intended victim it sets about fooling its immune system into thinking everything is just fine – nothing untoward at all. From the first day of infection, the parasite stimulates an over-activity of the gene FOXP3, that codes for a protein present in the regulatory lymphocyte that acts as a signal in the immune system, and effectively erases any immune response. The putative designer has used a feature of the immune system it designed to protect mammals like sheep, cattle and humans, and turned it against them, seeing the immune system it designed as a problem to be solved, if its flukes were going to do what it designed them to do.

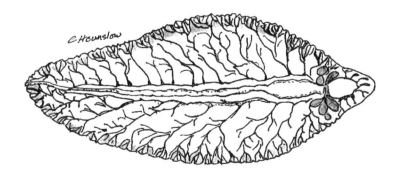

Figure 26 Liver fluke

After all the trouble Creationism's intelligent designer had gone to to design and create the liver fluke, the last thing it wanted was that troublesome immune system working the way it was designed to work and protect the sheep, cattle or human!

The University of Córdoba team explained the importance of this discovery because the drugs we currently use to control infestations by this nasty little platyhelminth are likely to become less effective as the worms are developing resistance to them. The putative intelligent designer is leaving nothing to chance.

Figure 27 Great-Spotted Cuckoo

Cuckoos and Other Parasitic Birds

European Cuckoos and Their Victims.

The European cuckoo, *Cuculus canorus,* is well known as a brood parasite on various species of passerine bird such as dunnocks, *Prunella modularis*, reed warblers, *Acrocephalus scirpaceus,* and meadow pipits, *Anthus pratensis*.

Figure 28 European Cuckoo, Cuculus canorus

The females normally parasitize the broods of the species that were their foster parents and have evolved egg mimicry. Although her eggs are larger than those of the host species, they closely resemble them in colour and markings. The evolutionary explanation for this mimicry is that it is the result of an evolutionary arms race in which the host learned to recognise the cuckoo's egg and remove it from the nest. Mimicking the markings of the host reduced this tendency. Parasitizing her foster parent species has produced a number of female gene lines which conserves this evolved mimicry.

However, those that parasitize the dunnock, which has plain pale blue eggs, are marked like those of the meadow pipit, leading some to believe that this is a recently parasitized species so the dunnock has not yet learned to recognise and remove the cuckoo's egg.

Some female cuckoos will also remove a host's egg from the nest and eat it, reducing the possibility that the host might be able to recognise an additional egg.

The cuckoo is a mimic in another sense too, in that they superficially resemble a bird of prey such as a sparrow hawk, complete with hooked beak, and barred breast markings, so they frighten sitting birds away long enough for the female cuckoo to lay her egg.

Cuckoo eggs have thicker shells than the host's eggs, giving them additional protection in the nest. They also hatch earlier than host eggs. If the cuckoo egg hatches before the host's eggs, the young cuckoo will wriggle under the egg and raise its featherless wings to hold the egg between its body and the edge of the nest, then, pushing upwards with its legs, it rolls the egg up and over the side of the nest. It does the same with any hatched chicks.

The gape of the chick is an exaggerated copy of that of the host's chick so inducing a powerful feeding reflex in the foster parents. The foster parents continue to feed the ever-hungry and rapidly growing chick even when it is much larger than they are and for longer than they would need to spend feeding their own chicks.

When mature, the young cuckoos follow their parents on migration to sub-Saharan Africa, the parents having migrated about a month earlier.

The evolutionary explanation for this parasitism is easy to understand, but, to avoid tedious repetition, I'll leave the reader to work out the

Creationist explanation without making the putative designer look malevolently sadistic – if that's at all possible.

Great-Spotted Cuckoos and Corvids.

The great-spotted cuckoo, *Clamator glandarius,* is a brood parasite on the carrion crow (*Corvus corone corone*) and its relative, the magpie (*Pica pica*). It occurs in Africa and the European Mediterranean coastal area.

Figure 29 Great-spotted Cuckoo, Clamator glandarius

The magpie is known to mob the great-spotted cuckoo and to remove its eggs from its nest. However, this behaviour is not seen in the carrion crow. Normally, this would be explained by the carrion crow being a recently parasitized species which hasn't yet had time to evolve these defensive behaviours. However, there is a possible alternative evolutionary explanation for this (lack of) behaviour in crows – there may be a degree of mutualism involved, as was shown by a team or researchers led by Daniela Canestrar of the University of Oviedo, Spain, from a study of great-spotted cuckoos in Northern Spain where two-thirds of carrion crows are parasitized by them (87).

The following is taken from a blog post I wrote at the time:

> Unlike many other avian parasites, great spotted cuckoo chicks do not evict host eggs and chicks from the nest but simply out-compete them in their demand for food from the adults. This often leads to reduced breeding success for the host species especially if food is scarce.
>
> The difference in nest location and construction between carrion crows and magpies is also significant in explaining the different responses of these two species to cuckoo parasites. Carrion crows nest in the tops of tall trees and construct an open nest, making the brood prone to predation by predators such as falcons and buzzards. Magpies, however, build a nest in a thicker cover and construct a roof over it, making it harder for flying predators to take the chicks.
>
> This difference in nest construction has created a different dynamic because it means the cuckoo chicks in carrion crow nests are also more liable to predation. This has produced evolutionary pressure for them to evolve a defence mechanism - they secrete a substance which repels predators. This in turn protects the host chicks in the same nest.
>
> Magpies, which don't benefit much from the anti-predator strategy of the parasite chicks because their chicks are not so heavily predated, and therefore suffer much more from

competition for resources with it, have had evolutionary pressure to evolve avoidance strategies.

Carrion crows, on the other hand, which can, in situations where predation is high, actually benefit from having a parasite chick in the nest, have had little evolutionary pressure to evolve avoidance strategies and might even be expected to evolve strategies to encourage it. However, given that predator numbers can fluctuate and there are years when predator numbers are low and so any potential benefit is not realised and loss due to competition can be the more significant effect, there may also be pressure acting in the reverse direction.

So, there is a dynamic of competing forces at work here which has probably reached an equilibrium at which the carrion crow does little or nothing to avoid parasitism because of the benefits it can get from it, but it has not evolved behaviours to encourage it because it also often suffers from it. Being dynamic of course means that it is relatively easy for a small change in one of the forces to push the dynamic in one direction or another, maybe over just a small part of the range, especially if the species involved is relatively sedentary.

And the evolution of the different strategies between magpies and carrion crows has its origins in the different nesting strategies adopted by the two species, which was itself probably produced by different responses to evolutionary pressures at some point in their evolutionary histories.

The historic forces which shaped evolution in the past may not even be present today so it can be difficult to see why something like different nesting behaviours in birds developed as related species diverged until we understand how natural selection operates (88).

Again, as with the previous section on the European cuckoo, I'll leave the reader to work out a possible Creationist explanation for all this, that doesn't involve malevolent intent or attempts to solve problems a putative designer created earlier.

Cowbirds and the Mafia Strategy

The cowbirds are a group of New World birds comprising six species each of which occupies a different geographical range over North, Central and South America and the Caribbean Islands. They are brood parasites on other birds.

Figure 30 Brown-headed cowbird, Molothrus ater

Females will watch the potential host until the nest is left unattended – usually in the afternoon – when she will dart in and quickly lay an egg. She will then continue to watch the nest. If the host removes the egg, as they often do, she will attack the nest and destroy it together with the host's eggs. This strategy is believed to reduce the evolutionary pressure on the host species to adopt the egg-removal strategy in the ongoing evolutionary arms race that inevitably follows brood parasitism.

This is known as the 'Mafia strategy' and was the subject of a 2007 paper by Jeffrey P. Hoover of the University of Illinois and Scott K. Robinson of the University of Florida, who observed this behaviour in the brown-headed cowbird, *Molothrus ater*. They found that 56% of nests where the cowbird egg was removed were depredated and only 20% of non-parasitized nests. Furthermore, 85% of replacement broods of depredated nests were also depredated. The cowbirds exacted a heavy toll on birds that removed their eggs from their nests. This 'Mafia strategy' results in an evolutionary equilibrium balancing the cost of rearing a cowbird young and having the brood destroyed.

Apart from the addition of thuggish, Mafiosi behaviour on the part of any designer who designed this reproductive strategy for cowbirds, it is difficult again to see this as anything other than the sadism of malevolent intent and the stupidity of an arms race where it tries to solve problems it designed earlier as solution to other problems it designed even earlier. And all to produce more cowbirds, when it had designed other birds to raise their broods themselves with a perfectly good reproductive strategy that doesn't involve parasitism on another species.

Cuckoo Finches and Arms Races.

The cuckoo finch or parasitic weaver, *Anomalospiza imberbis* of East Africa is an obligate brood parasite on other birds of the *Cisticolas* and *Prinias* families. As we have seen with the earlier brood parasites, this has resulted in the inevitable evolutionary arms race as the parasitized hosts evolve strategies to reduce the depredation and the cuckoo finch tries evolves ways to circumvent the defence strategies of its intended hosts.

The strategy adopted by the tawny-flanked prinia, *Prinia subflava* , is to change the colours of its eggs more quickly than the cuckoo finch can and to produce eggs in an array of colours with markings that the prinia recognises as its own, almost like a signature.

Figure 31 Cuckoo finch, Anomalospinza imberbis

The red-faced cisticola, *Cisticola erythrops*, on the other hand, while not varying the colour of its eggs has evolved to be better at spotting the cuckoo finch's eggs and removing them. Another species, the rattling cisticola, *Cisticola chiniana* which might be expected to be an ideal host for the cuckoo finch appears to have won the arms race and is not parasitized at all (89).

Parasitic Finches

The 19 species of African parasitic finch of the *Vidua* genus, also called indigobirds and wydahs are widespread in Africa and all are obligate brood parasites on species of grassfinch. Each species of *Vidua* specialises in parasitising a particular species of grassfinch. Unlike many brood parasites, female *Vidua* don't destroy the host's eggs and may lay several eggs in the same nest.

Grassfinch chicks are noted for their bright colouring and distinctive patterning and particularly the characteristic markings inside their open mouth and throat. These markings are believed to encourage the parents to feed them so those with the widest gape, normally the hungriest chicks, tend to be fed first. Chicks also induce the parents to feed them by characteristic begging postures and calls.

Because each species of indigobird or wydah specialises in parasitising one species of grassfinch, the parasite chicks closely mimic the grassfinch chicks in their markings, their begging postures and their calls. Mimicry sometimes exaggerates the host species' chick's markings and behaviour so the parasite chicks get preferential feeding. This astonishing degree of mimicry was shown by an international team of researchers led by Gabriel A. Jamie of Cambridge University Zoology Department, Cambridge, UK, and the University of Cape Town, South Africa, using computerised pattern-recognition to 'see' the patterns the way a bird sees them (90).

Since the parasite chicks are reared alongside the host species chicks, they become imprinted on the parent species and males tend to imitate the calls of the males of the grassfinch species that raised them, while females mate with males that imitate her host species' call and only lay eggs in that species nest.

This in turn reinforces the specialisation of each parasite species on one host species, which, biologists believe is how the different species arose in the first place. These preferences also act as barriers to hybridization in that females are unlikely to breed with males of related species which use the 'wrong' calls and hybrid chicks would have a mixture of colours and intermediate patterning instead of the highly accurate markings of a pure-bred chick, giving hybrid chicks a distinct disadvantage over the host chicks, so any tendency to hybridize would be quickly eliminated.

The evolutionary explanation for this is that mimicry will tend to improves since it gives more advantage to the parasite's chicks and imprinting ensures specialisation on a single host species. Barriers to hybridization then emerge because of the disadvantage any hybrids would have, so pushing the divergence into different species even further.

By contrast, because they reject this perfectly natural process, ID advocates have to attribute the whole thing to intentional design by a single intelligent supernatural entity so it complies with Biblical and Qur'anic creationism. This same supernatural entity also designed the host species which its parasitic species is obliged to parasitise on, apparently.

Parasitic Vertebrates

Most of the natural horrors we've encountered so far have been small or very small. The only vertebrates we've seen are the birds which are parasitic in the sense of brood parasites. They gain at the expense of another species, but unlike the viruses, bacteria, protozoans and worms of various types, do not live directly in or on another species.

But there is another diverse group of parasites that live by feeding on the blood of other animals, and not in the blood-sucking sense of the many blood-sucking insects and other arthropods. These haematophagous (blood-eating) horrors are all vertebrates.

The Vampire Fish.

Vandellia cirrhosa is a variety of catfish that can grow up to 40cm (16 inches) but most are considerably smaller. This fish is a native of the Amazon and its tributaries. It is supposedly notorious for swimming up a man's urethra as he urinates in a river, allegedly attracted by urine and swimming up the concentration gradient, eventually becoming lodged in the urethra aided by backward-pointing spines on the gill covers. However, these reports are dubious and remain uncorroborated. Some are unlikely in the extreme; for example, reports of the fish swimming up the stream of urine as the man urinates into water! There is however an account from 1891 by naturalist Paul Le Cointe, of V. cirrhosa being manually removed by him from a woman's vagina where it had become lodged, having entered head-first (91).

What V. cirrhosa **does** do, however, is feed on the blood of larger fish. It does this by entering under the gill flap and puncturing the fish ventral or dorsal aorta (the blood supply from the heart to the gills) and allowing the blood pressure to pump it into its stomach which can become engorged and extended to the degree that the blood can be seen through the abdominal wall. It is unlikely that the fish actually sucks blood from its victim. V. cirrhosa is reputedly able to engorge itself in 30 and 145 seconds.

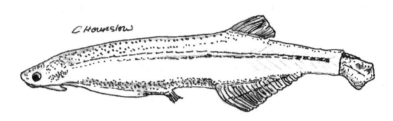

Figure 32 'Vampire fish' or candiru, Vandellia cirrhosa

When satiated or while waiting for a prey fish, V. cirrosa remains hidden in the river-bed sand and mud. It is thought that it detects its prey partly by detecting waste such as carbon dioxide being diffused through the gills.

Angler Fish and Sexual Parasitism

While on the subject of parasitic fish, I will turn now to the reproductive strategy of the angler fish and leave the reader to decide whether this bizarre strategy is the work of an intelligent, omnibenevolent designer. Some of this is taken from a blog post I wrote on the subject some years ago (92).

Angler fish, comprise some 16-18 families of bony fish. Some, such as the *Ceratiidae* show an extreme degree of sexual dimorphism and an unusual reproductive strategy probably to overcome a low population density so encounters between males and females would be rare.

When a male angler fish hatches, he can barely feed himself, if at all, but he can swim, has a powerful sense of smell with which he detects the merest whiff of a nearby female, and teeth. If there is no female close enough, he dies as he is incapable of fending for himself. Both he and unmated females have under-developed gonads.

Using this sense of smell, he locates a female and bites her skin, then secretes an enzyme which dissolves the skin of his own lips so they fuse with the female's body. Their circulations then merge so he gets all his nourishment from her. Needing no digestive system, this atrophies and is absorbed by the female, followed by his heart, brain and then the rest of his body, leaving just his gonads attached to her surface. This is one of the most extreme cases of sexual dimorphism known. Only then do their gonads mature.

Each female may have several sets of gonads from several males ready for when she needs them. When she is ready to spawn, she produces a hormone which causes 'her' gonads to produce sperm at the moment she lays her eggs. She never has any need to find a mate. It is like being hermaphrodite, yet has all the advantages of sexual reproduction and, in the case of multiple parasitism, of having several mates to ensure a good genetic mix.

Recently, researchers at the Max Planck Institute of Immunobiology and Epigenetics in Freiburg, Germany found that to make this possible, angler fish have a substantially reorganised immune system which supresses any tendency of the female to reject the male, or vice versa, yet which leaves them able to resist infections by other organisms. Although the precise details of this replacement immune system remain unknown, what they found was that angler fish have lost most of the genes that normally produce the histocompatibility antigens that are normally found on the surface of cells and which act as signals to the immune system when a cell is under attack from a bacterium or virus. They have also lost the function of the T-cells that normally kill infected cells as the body mobilises against harmful pathogens (93).

In other words, the angler fish has traded immunity for sex and the immune system a putative intelligent designer designed for it has had to be substantially redesigned – and all because the population of angler fish was too low for males to find females.

Boring Lampreys

Lampreys are considered to be primitive vertebrate chordates, descendants of a stem group which includes hagfish and a sister group of the Gnathostomata (jawed vertebrates) which gave rise to the bony fish and the Chondrichthyes, which gave rise to the cartilaginous sharks, skates and rays.

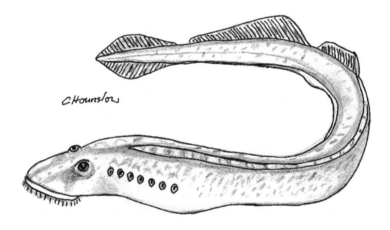

Figure 33 Marine lamprey, Petromyzon marinus

Lampreys live in both fresh water and sea water with some fresh-water species returning to the sea to breed. Superficially, they resemble eels but they have seven gill-slits and a suctorial disc in place of a mouth. The freshwater species do not feed as adults, relying on fat stores laid down as filter-feeding larva or ammocoetes. The marine species, however are predatory on fish.

Having attached themselves to the body of the fish with the suctorial disc, aided by rows of teeth, they then rasp a hole in the fish until they reach its body fluids. Some species eat the flesh as well. They may even enter the body of the fish itself and eat the internal organs.

Having laid eggs which are fertilised externally, the adults die. The eggs hatch into the larval form, or ammocoetes. In the case of freshwater lampreys living in rivers, these drift down with the current to estuarine mud and silt where they partially bury themselves tail first and feed by filtration.

For reasons which are not understood, lampreys have the largest number of chromosomes for any vertebrate (164-174) in contrast to the human 46. This conflicts with the Creationist dogma that evolution leads to greater complexity and greater complexity requires more genetic information, therefore a larger genome. Lampreys are amongst the least complex of the vertebrates. In 2018, a team of scientists led by Jeramiah J. Smith of Kentucky University, showed that this high number of chromosomes had come about by gene duplication early in their evolution (94). They suggested that this could have played a role in the evolution of the vertebrates by providing plenty of opportunity for beneficial mutation in the duplicated genes. This contradicts Creationist dogma that all mutations are harmful and that no new genetic information can arise by mutation.

Marine species having been accidentally introduced into the North American Great Lakes, have adapted to fresh water, breeding in the feeder streams, and are becoming a serious pest, predating on fish.

Vampire Bats.

The three species of vampire bat are all native to South and Central America. They feed exclusively on the blood of warm-blooded animals. Vampire bats are the only truly parasitic, haematophagous mammals.

The common vampire bat, *Desmodus rotundus,* feeds almost exclusively on mammals and approaches its intended victim at night while they sleep. It uses heat sensors in its nose to locate blood vessels near the surface, then it cuts a flap in a soft part of its skin with razor-sharp teeth, then licks up the blood with its long tongue. Anticoagulants in the bat's saliva prevent the blood from clotting.

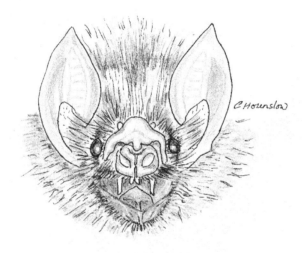

Figure 34 Common vampire bat, Desmodus rotundus

The common vampire bat, *Desmodus rotundus,* feeds almost exclusively on mammals and approaches its intended victim at night while they sleep. It uses heat sensors in its nose to locate blood vessels near the surface, then it cuts a flap in a soft part of its skin with razor-sharp teeth, then licks up the blood with its long tongue. Anticoagulants in the bat's saliva prevent the blood from clotting.

These bats are highly social, with one male dominating a large number of females. They mutually groom and will feed regurgitated blood to members of the colony that have not fed. Like the other vampire bats, the common vampire is highly mobile on the ground, able to run and jump, using its folded wings, with well-developed clawed thumbs complete with pads, as forelimbs. Because it feeds on livestock and carried rabies, it is considered a pest. It will bite humans while they sleep. Most cases of rabies in humans in the USA are caused by vampire bat bites.

The hairy-legged vampire bat, *Diphylla ecaudata,* comprises two sub-species, each with a different range over South and Central America. It feeds mostly on the blood of birds, including domesticated birds and has been known to bite humans. It is superficially similar to the common vampire bat but has shorter thumbs which lack the pads of the common vampire.

It will also share regurgitated blood with other members of its species, especially its close kin and mothers supplement their young's milk diet with regurgitated blood. Like the common vampire, this bat can also carry and transmit rabies.

The third vampire bat is the white-winged vampire bat, *Diaemus youngi,* found in Central and northern South America and the Caribbean islands on Trinidad, Tobago and Margarita. They feed on both birds and mammals, especially domestic goats and cattle and sometimes humans. Their thumbs are shorter than those of the common vampire so they are less agile on the ground. They also, rarely, carry rabies.

The Vampire Ground Finch.

The vampire ground finch, *Geospiza septentrionalis,* is a native of two of the Galapagos Islands, Wolf and Darwin, where, especially when food is short, it feeds on the blood of birds, especially the Nazca and blue-footed boobies, *Sula granti* and *Sula nebouxii,* respectively by pecking at their feet until they bleed. This feeding behaviour is believed to have evolved from pecking at the plumage to remove parasites. Strangely, the boobies do not offer much resistance to this pecking.

The vampire ground finch also parasitizes other species by taking and eating eggs, having learned to roll the eggs onto rocks until they break.

Figure 35 Vampire ground finch, Geospiza septentrionalis (female)

As an interesting example of how birds diversify by call. On Darwin, vampire ground finches have a buzzing song and on Wolf, a lilting song (95).

The Boring Snub-Nosed Eel.

The snub-nosed eel, *Simenchelys parasitica*, is endemic to the North Atlantic and the Pacific Oceans where it lives close to the ocean floor, mostly scavenging on the decaying corpses of marine creature, much in the way hagfish do, using its powerful jaws and teeth to tear off lumps of dead flesh. But it also has another option for feeding. It also feeds on living fish, rather like the lampreys we met earlier.

It attaches itself to the side of the fish and bores its way through into the body cavity, where it can live for some time. Two juvenile females were even found within the heart of a mako shark where they had been living on blood. The evidence was that they had been there for some time.

The American biologist, Spencer Fullerton Baird, of the Smithsonian Institute reported that snub-nosed eels can be, "Not unfrequently found nestling along the backbone of the halibut and cod" and that they will also enter the abdominal cavities of netted shad and eat their eggs, all within a few minutes (96).

Figure 36 Snub-nosed eel, Simenchelys parasitica

The inventive, maniacal genius of a designer who comes up with these designs is truly astonishing, as is the mentality of those who believe it should be regarded as the inspiration for human morals and worthy of praise.

The Cookiecutter Shark.

The small, cookiecutter shark, *Isistius brasiliensis,* is widespread in the warmer oceans. It is between 45 cm and 56 cm (16.5-22 inches) long and has an ingenious method of feeding, not unlike that of a very large lamprey. It is indiscriminate, attacking large aquatic mammals and fish and has been known to attack humans. It attaches itself to the animal with its suctorial mouth, then, anchored by its upper teeth, and using its razor-sharp lower teeth that resemble a chain saw, it cuts out a circular lump of flesh by rotating round and vibrating its lower teeth rather like an electric carving-knife, leaving a gaping wound about 5 cm (2 inches) across and 7 cm (2.8 inches) deep, that resembles what a cookie cutter might make.

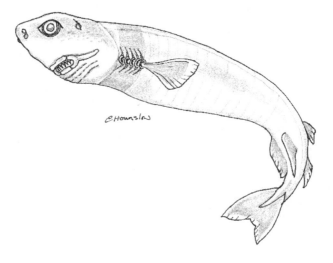

Figure 37 Cookiecutter shark, Isistius brasiliensis

Its prey includes most species of whale and dolphin, seals, manatees and dugongs as well as large fish such as tuna and other sharks. One beached whale was found to have hundreds of fresh wounds and healed scars of cookie-cutter bites. Swimmers in the seas around Hawaii have been attacked (97).

Like all sharks, *I. brasiliensis* sheds its teeth regularly but, to retain its eating ability, it sheds its lower teeth all together then eats them to recycle the calcium in them. They are replaced immediately by the next set which have been developing behind them.

I. brasiliensis also hunts conventionally and will ingest prey such as squid whole, even though they may be almost as large as itself.

Blood-Sucking Oxpeckers.

The familiar oxpeckers used to be thought of as examples of mutualism or symbiosis as the birds cleaned the large mammals of ticks and other ectoparasites. However, new evidence shows that they should be regarded as ectoparasites in their own right.

Figure 38 Yellow-billed oxpecker, B.a.africanus, on giraffe

They are comprised of two closely-related species, the yellow-billed and red-billed oxpecker, *Buphagus africanus* and *B. erythrorynchus* respectively. They are sometime regarded as subspecies of *B. africanus, B.a. africanus* and *B.a. erythrorynchus*. They have almost identical *modus operandi*. They both settle on large mammals and, if there are not enough ticks or other ectoparasites, will open up a closed wound or create a new one and feed on blood (98).

Figure 39 Toothwort

Parasitic Plants

So far, we've looked mostly at the parasitic animals, from worms, insects, birds and mammals and other vertebrates. It is time now to look at some of the parasitic plants. These are especially difficult for ID advocates to explain as there are so many examples of plants living perfectly independent lives and many of the parasites have relatives that do so. The conclusion, if we accept the ID explanation for biodiversity, is that the intelligent designer wanted to make life more difficult for the perfectly regular plants that suffer from parasitism.

In July 2020 a team of researchers at the Bioscience and Biotechnology Center, Nagoya University, Japan showed that parasitic plants produce an enzyme, β-1,4-glucanase, to help attach themselves to the roots or stems of their host plants (99). To an ID advocate, the explanation for this has to be that the parasitic plants have been intelligently designed to produce this enzyme to make it easier for them to parasitise the plants this putative designer also designed! They consider this a better explanation than that provided by evolution by natural selection.

To a plant, another plant is simply a potential niche; a resource that is available to be exploited should suitable mutations arise to make it possible. If that produces more copies of the plant then the mutation will increase in the species gene-pool. Again, as with the animal parasites, there is no emotion or moral judgement evolved. If it gives an advantage it works; if it creates a disadvantage, it fails and gets eliminated from the gene pool. No plan and no direction.

As a youngster in a small hamlet in North Oxfordshire, one of our haunts was a small deciduous wood which, before the ravages of Dutch Elm disease, was almost entirely elm with a few other trees such as hazel, holly and sycamore. In one small patch, in Spring, we regularly found strange-looking pinkish-white plants growing straight out of the ground. We called them "dead man's fingers". They were almost entirely pinkish-white with no sign of green. We debated endlessly over whether they were plants or fungi – or really the result of a dead body buried in the wood.

I now know these to be the Toothwort, *Lathraea squamaria*, a plant parasite on the roots of hazel, elder and elm, not the gruesome indication of a buried murder-victim as legend and children's tales had it.

Toothwort.

The toothwort, Lathraea squamaria is an obligate, heterotrophic parasite, as evidenced by its lack of chlorophyll which means it is entirely dependent on its host for all it nutrients. They give nothing back to their host. In addition to the flowers that appear in Spring, *L squamaria* also produces underground 'flowers' which fail to open and are self-pollinating within the closed flower.

In 2016 a team of researchers from Lomonosov Moscow State University, Moscow, Russia, showed that *L. squamaria* has non-functional chloroplasts (plastids) but that the genes necessary for photosynthesis have been broken. Only one of the 22 genes was missing – *petL* – but the others had mutations that made them non-functional. They suggested that this showed little selection pressure to remove deleterious mutations, indicating that transition to holoparasitism was relatively recent. (100)

Of course, ID advocates would presumably argue that these missing and broken genes could not have given *L. squamaria* an advantage because Creationist dogma states that all mutations are harmful and that evolution always results in **more** genetic information, not less. In their world, intelligent magic is responsible for redesigning *L. squamaria* so it can live off elm tree roots without giving anything back, but where is the intelligence or future planning in giving *L. squamaria* the genes for photosynthesis then breaking them?

Figure 40 Toothwort, Lathraea squamaria

L. squamaria is a member of the *Orobanchaceae* family of parasitic plants that includes the broomrapes. These plants are believed to have lost the ability to make chlorophyll on at least two occasions. One member of the family, *Orobanche ramosa* is a major parasite on a whole range of agricultural crops, including, tobacco, tomato, potato and Brassicas, including oilseed rape.

Hemp Broomrape.
The hemp broomrape, *Orobanche ramosa* sends up flower-bearing stems from a thick root. The stems are devoid of leaves and chlorophyll so the plant is entirely dependent on the plant it attaches too. This reduces crop yield and, in severe infestations, the entire crop can be lost. It can be spread by contaminated seed or soil and is difficult to eradicate once established.

A single plant can produce up to a quarter of a million seeds which can lie dormant in the soil for many years, until stimulated to germinate by chemicals given off by plant roots. This simple mechanism ensures that when the seeds germinate, they have plant roots nearby to attach to and begin taking water and nutrients from them, otherwise they just lay dormant in the soil and wait.

Once established, reusing the contaminated land again within 15-20 years can result in rapid re-establishment of the weed. The germinating seed sends out thin, rootlike structures which attach themselves to nearby plant roots, and draw all their nutrients and water from the host plant (101).

Figure 41 Hemp Broomrape, Orobanche ramosa

In some parts of the world, tomato crops can be reduced by 30-100%. In Germany, rapeseed yield was reduced by between 58% and 70% in one reported case. O. ramosa is spreading into western France where it is threatening rapeseed, tobacco and hemp production (101).

Creationism's intelligent designer appears to be waging a war on human agriculture too!

Witchweeds.

Witchweeds or *Striga* is a family of parasitic plants of the *Orobanchaceae* family. They are hemiparasitic, obligate parasites on the roots of plants, which are essential for their initial germination and development, after which they are capable of independent existence. There may be over forty different species.

Three species in particular, *Striga asiatica*, *S. gesnerioides*, and *S. hermonthica,* can devastate the crops of subsistence farmers. The crops most at risk include maize, millet, sorghum, sugarcane, rice and legumes. Because of the heavy toll on the water and nutrients, infestation can resemble the results of nutrient deficiency and drought. Entire crops can be lost.

The plants can each produce between 90,000 and 500,000 seeds which, like those of *O. ramosa* can lie dormant in the soil for over 10 years. As with *O. ramosa,* the seeds of *Striga* germinate in response to chemicals produced by plant roots.

The mechanism by which *Striga* find and attach themselves to the roots of their hosts is ingenious. Having been initially stimulated to germinate, the seed sends out a root-like structure that probes the soil, releasing an oxidizing enzyme that dissolves plant roots causing them to produce quinone. When the concentration of quinone reaches a threshold, a haustorium develops which then follows the quinone concentration gradient until it makes contact with the root. At that point the cells in the haustorium multiply rapidly to form a wedge-like structure and the cells elongate to force the haustorium into the root. Finger-like projections, called oscula then grow from the haustorium to make contact with the conductive fibres in the root and begin drawing water and nutrients into the seed, which then produces cotyledons to grow upwards to become a new plant.

Figure 42 Small Witchweed, Striga bilabiata bilabiata

In Africa, witchweed affects 40% of arable land and costs some $30 billion in lost crops per annum. A little gift from the intelligent designer to a continent already suffering from drought and famine.

Giant-flowered *Rafflesia.*

The *Rafflesia* are a genus of parasitic plants found in Southeast Asia. They parasitize vines of the *Tetrastigma* genus (relatives of the grape vine and the Virginia creeper. They are obligate parasites and have entirely lost their ability to produce chlorophyll or to perform photosynthesis. There are an estimated 28 different species all with similar life-styles.

They are of enormous botanical interest (102) but have little economic impact. They are characterised by their only visible manifestation – the enormous *Rafflesia* or corpse flower, which smells of rotting flesh. Some specimens can be up to 4 feet (120 cm) across. The smell attracts pollinating insects which normally feed on carrion.

They get all their nutritional requirements from their host by a system of haustoria which infiltrate the plant in close association with its conductive tissues. Their botanical interest comes from two unusual features:

1. They have lost the entire genome of their chloroplasts rendering them non-functional. The speculation is that this may have been the result of their parasitic life-style sometime during their evolutionary history (103).

2. They appear to have gained some genes from their host in a case of horizontal gene transfer. This is highly unusual for a higher organism. The speculation is that this may have given *Rafflesia* some protection against an immune response by its host. Analysis of these received genes also shows genes from other former hosts as a record of the evolution of *Rafflesia*. (102).

Figure 43 Rafflesia tuan-mudae

Creationism's intelligent designer seems to have been having some fun with Rafflesia, making it look for all the world like it evolved by a natural process, and the idea of making a flower smell like rotting flesh to get pollinated is bizarre, almost beyond belief.

The *Balanophora.*

This group of parasitic flowering plants is found in forests in tropical and temperate Asia, the East Indies, the Pacific islands, Madagascar and tropical Africa. There are about twenty known species, some of which are extremely local, like *Balanophora coralliformis* which is only found on Mount Mingan on the island of Luzon in the Philippines.

They are all obligate root parasites, having no chlorophyll of their own and leaves reduced to little more than tiny scales on the stems of the above-ground flowering inflorescence, which, in most species, resembles a puffball except that the head is covered in tiny female flowers and is surrounded at its base by a relatively few male flowers. *B. coralliformis,* known only from about 50 specimens has a branched, coral-like inflorescence.

Probably to attract pollinating insects and other arthropods, like the *Rafflesias*, they emit an odour like rotting flesh or other savoury smells like that of mice. They produce the smallest known flowers of any flowering plant and the 100,000 to 1 million fruits on each inflorescence are minuscule, dry and born close to fleshy, coloured bracts which act as attractants for the birds which are the main dispersal agents, as was shown recently by Kenji Suetsugu from Kobe University, Japan, in regard to *B. yakushimensis* (104). The birds get their nutrient 'reward' from the bracts, not the fruit.

I will leave the reader to try to find a sensible reason why an intelligent designer would come up with this prolifically wasteful method for making more *Balanophora*.

The Dodders.

Dodder, or *Cuscuta* is a genus within the Convolvulus (*Convolvulaceae* - morning glory, bindweed, etc.) family, of 201 parasitic plants. They are parasitic on a wide range of agriculturally or horticulturally important crops such as alfalfa, clover, potato, tomato, dahlias, petunias and even ivy.

Dodder seeds germinate on or close to the surface of the soil and to begin with is independent. However, it must find a host plant to attach to within 7-10 days or it will die. Having attached itself to its host, it is then entirely dependent on it for all nutritional requirements. Unlike the *Orobanchaceae* we met earlier, dodder dos not attach to its hosts roots but winds itself round their stems, like its climbing relatives, the bindweeds. It then inserts haustoria into the stem of its host and attaches to the conductive tissues. The root in the soil then dies.

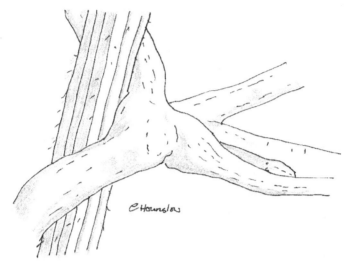

Figure 44 Dodder, Cuscuta spp.

As it grows, dodder may attach to other nearby plants of the same or different species and can then act as a bridge to facilitate transfer of diseases between plants. Dodder also reduces the host's ability to resist viruses, probably by suppressing its immune responses. In 2006 it was shown that dodder can locate its potential host and distinguish between them using chemicals given off by their hosts (105).

One of the problems leafless perennial plants like the dodders have is in timing their flowering. Normally this is coordinated by the leaves of flowering plants which produce a protein hormone called 'Flowering Locus T' (FT) to signal the plant to start to flower. By tagging FT with a green fluorescent protein, a team from the Max Planck Institute for Chemical Ecology's partner group in the Kunming Institute of Botany at the Chinese Academy of Sciences, showed that *C. australis* 'eavesdrops' on its host and so uses the host's leaves to signal the best time to flower (106). How brilliant is that?

However, there are signs that some hosts may be staging a fightback. Tomatoes have been shown to activate two metabolic pathways in response to attack by *C. pentagona,* the five-angled dodder, native to

North America. They produce complex volatile chemicals which may interfere with the dodder's ability to detect a potential host. Also, the fine hair-like structures (trichomes) on the stems of tomatoes may inhibit dodder from attaching itself (107).

Has Creationism's intelligent designer started another arms race with itself?

Mistletoe.

Mistletoe, *Viscum album*, one of a number of parasitic plants of the *Viscum* genus, is a very familiar parasitic plant on several different tree species in UK. It seems to be increasing. As a child in North Oxfordshire, I only ever saw one mistletoe plant growing on an apple tree in a friend's garden; it is now very common in and around Oxford and South Oxfordshire with some trees so heavily infected that they are close to being killed by it.

The familiar (in UK) mistletoe, *Viscum album*, is an obligate, hemiparasitic plant that actively photosynthesises with green leaves and stems. There are many legends and superstitions surrounding it probably because of the 'magic' way it grows from another plant and not from the ground. It was supposedly sacred to Druidism. The tradition of kissing beneath a sprig of mistletoe at Christmas is probably a pagan fertility rite. There is a belief that mistletoe is banned from being taken into English churches but this is not a formal prohibition, merely a folk memory of the pagan rituals associated with both mistletoe and Yule at mid-Winter.

Mistletoe is normally spread by birds that eat the berries and for whom they are important mid-Winter foods. One bird in particular, the blackcap, which has recently started migrating to UK from Central Europe in the Winter to give us a Winter population (108) is also becoming an important vector for spreading mistletoe.

Figure 45 Mistletoe, Viscum album

Once mistletoe becomes established on a tree it will attract birds to eat the berries. These pass very quickly through the digestive tract to be deposited on a branch where they stick with what remains of their highly adhesive coat. Other birds regurgitate the seeds or wipe them off their beaks onto a handy branch, so the tree can quickly become heavily infected, even losing whole limbs.

The Asian species, *V. continuum,* completely replaces the crown of the tree, leaving only the trunk on which it is dependent.

When the seeds of *V. album* germinate, the embryo plant is independent until it can send a hypocotyl into the bark of the tree. Each seed may produce two or four embryo plants, each of which can develop into a full plant to produce a bunch sprouting from the same location. Once the hypocotyl has penetrated the bark it sends a haustorium into the tissues of the tree to attach to the vascular tissues. At that point it becomes dependent on the tree for nutrients and water but continues to make its own sugars by photosynthesis.

The Australian Christmas Tree.

The Australian Christmas Tree, *Nuytsia floribunda*, so called because it flowers at the Australian Christmas – in mid-Summer - is a hemiparasitic distant relative of mistletoe. However, unlike most plant parasites it does not grow directly on its host plant(s) but taps into their root systems underground. In fact, it solves the problem of water-shortage in the Australian desert in an unusual way – by stealing the water-reserves and sources of other plants.

N. floribunda is fully photosynthesising complete with green leaves, but underground it has an extensive network of roots which, when they come into contact with the roots of other plants, send haustoria into them, tapping into their vascular tissues and extracting water and nutrients. Almost all species are parasitized. They will even send haustoria into underground cables! (109)

The root system gives rise to suckers, so an extensive grove of *N. floribunda* can develop all extracting water and nutrients from surrounding plants. Between October and January, it bears masses of vivid yellow-orange flowers on inflorescences that can be a metre (3 feet) long, giving it its alternative name – the fire tree.

It is considered semi-sacred by the Nyungar people of Southwest Australia who obtain a sweet gum as a delicacy from it, so it has a protected status. So popular is the sweet gum with children that they are told stories of avenging spirits coming in the night if they eat too much of it.

If you're looking for inspirational morals from this putative intelligent designer, look again, unless you see taking from your neighbours in times of need as moral.

Love Vines and Parasitic Gall Wasps

There are some 13,000 different species of parasitic gall wasps that lay their eggs in plant tissues. The grubs then produce substances that induce the plant to produce a gall in which it can grow and feed before exiting the gall to pupate.

Many of these wasps have a two-phase life cycle in which one phase is sexual and the other is asexual, all-female, reproducing parthenogenically to produce the next (sexual) phase. The sexual phase normally produces galls on the stems and leaves of plants while the asexual phase parasitizes the roots.

But these wasps and their parasitic life-style are not the example of malevolence here. In fact, most gall wasps appear to do little harm to the host species. Where the malevolence comes in is in the relationship between several of these wasps and a parasitic plant – the love vine, *Cassytha filiformis*, which is widely distributed throughout the tropics. A team of researcher at Department of BioSciences, Rice University, Houston, Texas, led by Scott P. Egan, found that the Florida subspecies of *C. filiformis* appears to target the galls of several gall-forming cynipid wasps that normally parasitize the sand oak, *Quercus geminate,* especially *Belonocnema treatae*, which makes single-chambered galls on the underside of the leaves of the host oak (110).

Like other parasitic plants, *C. filiformis* creates an attachment organ, the haustorium, which penetrates the host tissues and extracts moisture and nutrients. When the scientists examined the galls with haustoria attached, they found that not only had moisture been drawn from the plant but also from the wasp grubs inside it. Almost half the parasitized galls had dead and desiccated wasp grubs in them, as opposed to only two percent of non-parasitized galls. A parasitic plant that appears to seek out and kill wasps inside galls! The wasp grubs, of course, are trapped in their gall and incapable of defending themselves as their body is sucked dry.

Not only that but the same team found that the galls parasitised by the love vine tended to be larger on average than the non-parasitized galls. It seems unlikely that the vine could select the largest galls – which would mean measuring them and the others to determine which to select and which to reject, so the most likely explanation is that the haustoria somehow make the galls grow larger.

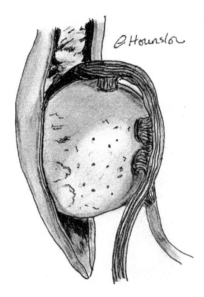

Figure 46 Wasp Gall on Oak Leaf and C. Filiformis Haustoria

At the expense of labouring the point, ID advocates must regard this as the deliberate design of a creative intelligence and, to conform to monotheist dogma, the work of the same designer who designed the gall wasps.

The researchers also found that a less-common species of wasp, *Callirhytis quercusbatatoides*, which forms multichambered galls on the stems of sand oaks, is also attacked by *C. filiformis,* as are the galls of four other species of wasp and one species of gall-forming fly, *Arnoldiola atra.*

This is believed to be the first known instance of a plant actively parasitizing an insect. The Rice University press release announcing this discovery (111) was accompanied by a YouTube video (112).

Murdering Hawkweeds.

My last example in this chapter is not really a parasite in the true sense of the word, but can you get more parasitic than murdering your neighbours to take over their land? These are the hawkweeds of the *Hieracium* genus.

This family of plants are closely related to the dandelion, chicory and sow thistle group of the large family which includes sunflowers and asters. In places, they have become serious invasive weeds in agricultural fields and meadows, as they can quickly replace other species, reducing biodiversity and making it unfit for grazing. The genus gets its name from the ancient Greeks who believed hawks (hierax in Greek) ate the sap of hawkweeds or sprinkled it in their eyes to clear their vision.

Some species of hawkweed reproduce sexually but most produce seeds which are clones of the parent plant. This process is known as apomixis or agamospermy. Because the plants which grow from these seeds are almost genetically identical to their parent plant, a colony can be regarded as technically a distinct species since they do not interbreed with other colonies. These are sometimes referred to as micro-species.

However, the flowers of hawkweeds still produce pollen!

So, what is the pollen for? In 2009, researchers showed that the pollen of *Hieracium* kills the developing embryos of neighbouring plants when it comes into contact with their flowers (113) by releasing phenyl acetic acid. This prevents their neighbour's seeds from developing and so leaves more space for their own offspring.

The evolutionary explanation for this is straightforward and obvious – as a reproductive strategy it ensures greater reproductive success – but ID advocates are obliged to explain this as the deliberate design of the same intelligent designer who also designed all the other species that hawkweeds murder in their quest for space, like a genocidal invading army looking for lebensraum.

Supposedly, having designed the reproductive systems of the sexually-reproducing species, complete with their pollen-dispersal and receiving apparatus, it saw this as an opportunity to insert hawkweed pollen, with its lethal payload, into the flowers to kill their developing embryo plants.

Miscellaneous Malevolence

So far, we've seen how Creationism's putative designer seems to have designed a whole army of parasitic and pathogenic organism, from the very small to the large, from viruses and single-celled organisms to fungi, parasitic vertebrates and plants to make life difficult for the other living things it supposedly designed. I will turn now to instances where this putative designer has designed systems, organs and processes for some species that would have benefitted other species such as humans but, for reasons which, if we accept the ID view, can only be regarded as either wilful neglect or mendacity, it failed to give to others of its creation.

Elephants Don't Get Cancer.

Elephants are almost unique amongst mammals in that they rarely get cancers. This appears to contradict Peto's Paradox (114) which states that, since cancers are mostly caused by faults in DNA replication during cell division, animals with more cells or which live longer should have more cancers than those with fewer. Elephants are very large and tend to be long-lived yet rarely get cancers.

This paradox was partly resolved by the discovery of a gene, TP53, which produces a protein P53, which regulates cell division and stops cells proliferating too rapidly. This acts as an anti-cancer gene but the problem is, it is not very effective, as the incidence of cancers in animals like humans attests.

So, why don't elephants have the same rate of cancers as other mammals, when they too have this not-very-effective TP53 gene?

The answer was discovered in 2016 by researchers from the University of Chicago, who found that elephants have multiple copies of the TP53

133

gene, probably produced by gene duplication during their evolution to large size! So much for the Creationist dogmas that no new information can arise by mutation, and mutations are always harmful.

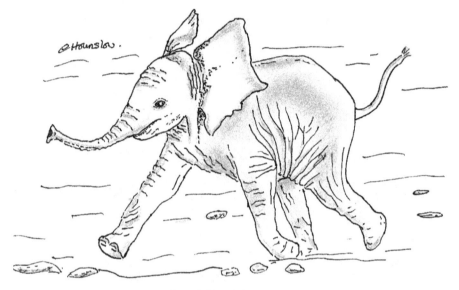

Figure 47 African elephant calf

What they discovered was that the increased activity of these genes means DNA damage either get repaired much more quickly than with other mammals and that, if it can't be repaired, cell apoptosis (cell death and destruction) is much more efficient, so either the cell fails to develop into a cancer because the DNA fault is repaired, or it is rapidly destroyed (115).

The ID view must be that this was deliberately designed by the same intelligent, omnibenevolent designer as the one who designed humans and all the other animals that suffer from cancer, yet, having designed a solution to the cancer (which it also designed) it failed to provide other species with it. The conclusion can only be that, for some reason, it decided to exempt elephants from its cancers but wanted other animals, including humans to die from them, and, having designed a singularly ineffective anti-cancer gene, it left it at that.

Bats' Superior Immune System.

We've already seen, in relation to SARS-CoV-2 (coronavirus) how bats appear to be immune to these viruses, so can act as incubators for new strains to arise, such as the one currently killing hundreds of thousands of people world-wide and wrecking economies (50).

Figure 48 Brown Long-eared bat, Plecotus auritus

Now we learn that bats have an immune system that is much more effective than that of almost all other mammals, including humans (51). Although it is very good at recognising foreign RNA, in viruses for example, it is not so good at recognising foreign DNA, which dampens the inflammatory response that does so much damage, especially in humans infected with SARS-CoV-2. The genes that control those responses have probably (according to evolutionary biologists) been purified by natural selection because bats live in a wide range of ecological niches and exploit multiple prey species, including many different species of insects.

According to ID advocates, however, evolution by natural selection cannot account for these differences – they must have been intelligently designed.

So, as with the elephant's immunity to cancer, Creationism's putative designer chose to give all but bats an inferior immune system, the better to allow the pathogens it designed to do what they were designed to do – make us and other animals sick.

Bird's Superior Eyes.

Despite the claims of spectacle-wearing Creationists like Ken Ham, the human eye is far from perfect. Their putative creator has created far better eyes for birds, for example.

Birds' eyes are more similar to those of their reptilian ancestors than they are to those of mammals. They have a lens which can be changed in shape more so and more quickly than the lens of mammalian eyes, giving much greater depth of vision. The bird's eyes are not spherical but are flattened from front to back, so more of the visual field is in focus than that of mammals like humans. A bird's eye is relatively large in proportion to their body – an ostrich's eye is twice the size of a human eye, for example. The trade-off is that birds' eyes are fixed in the skull so only a few birds can move their eyes within their sockets.

Some birds, like terns, gulls, albatrosses and other seabirds have red or yellow oil droplets in their colour receptors to improve vision in hazy conditions. Raptors such as eagles and peregrine falcons have a very high density of receptors in their retina to give a much higher visual acuity than most mammals. Humans have about 200,000 receptors per square millimetre in their retinas whereas the house sparrow has about 400,000 and the common buzzard about 1 million. Unlike humans, but like some mammals such as cats, birds clean their eye not with their eyelids but with a 'nictitating membrane' that wipes the eye from the inner corner, like a windscreen wiper. A peregrine falcon, stooping at speeds approaching 250 mph, uses the nictitating membrane to protect, lubricate and clean its eyes, while keeping visual contact with its target, like a pair of built-in goggles. This thin membrane is translucent, so the bird can still see. Humans have the vestige of this membrane in the form of the plica semilunaris – the pink fold of skin in the inner corner of the eye.

Figure 49 Peregrine falcon, Falco peregrinus

As with all vertebrate eyes, the retina has a blind spot because the nerves and blood vessels that supply the retina must pass through it. Bird eyes also have a structure whose function is not fully understood – the pecten oculi – which may have a role in reducing dazzle and detecting movement by casting a shadow on the retina. It may also have a role in providing nutrients to the rods and cones in the retina.

Each of the colour-detecting cones in both bird and reptilian eyes have a coloured oil droplet. This filters the light reaching the light-sensing part of the cone so improving colour vision. Birds have six different types of oil droplet increasing the number of different colours that birds can see. They can even see UV light!

Recently, a team of researchers led by Mary Caswell Stoddard, of Princeton University's Department of Ecology and Evolutionary Biology, with colleagues from the University of British Columbia (UBC), Harvard University, University of Maryland and the Rocky Mountain Biological Laboratory (RMBL) showed that wild broad-tailed hummingbirds, *Selasphorus platycercus,* can perceive several 'non-spectral' colours against humans ability to perceive only one – purple. This ability gives them the ability to identify the flowers on which they feed in much the same way that bees do (116) (117). A non-spectral colour is one which doesn't appear in the rainbow. Purple is a combination of red and blue, for example, which means we see two widely-separated wave-lengths in combination. Because *S. platycerus* can detect UV light, they can also see several colours in combination with UV.

Humans cannot resolve flickering light above 50 pulses per second, above which movement appears to be smooth (hence we see smooth movement in films shown at 50 or more frames per second. Birds, however can distinguish up to 100 pulses per second. This enables a sharp image to be seen at speeds at which, to a human, near objects would appear as a blur, so flying at speed through woods and branches of trees is possible.

Apparently, it was beyond the wit of Creationism's intelligent designer to use this superior eye design for humans. Instead, we have eyes that deteriorate as we age, so we need the services of science and opticians to provide us with the spectacles that give the lie to claims of perfection in design.

Birds' Superior Respiratory System.

The following is based on a blog post I wrote in 2014 (118):

> *It is considered a real achievement and a feat of endurance for a super-fit human, after prolonged training, to climb Mount*

Everest without oxygen tanks and breathing masks. It was first done as recently as 1978. Everest is a mere 8,848 Metres high. In 1975, a jet flying at a height of 11,264 Metres sucked a griffon vulture into its jet engine.

The problem with mammalian lungs like ours is that we draw in fresh air, which mixes with the stale air left in our lungs, trachea, bronchioles and alveoli, so it is already partly stale by the time it gets to the alveoli where gas exchange takes place. We then breathe out as much as we breathed in, leaving a substantial amount behind to contaminate the next intake. Physiologists refer to this as the 'dead space'.

Because of this, we need particularly large bronchioles to shift enough oxygen in and get rid of enough carbon dioxide and, with particularly high demands such as sprinting we can literally run out of breath. Our muscles don't get enough oxygen, use up their own reserves, and burn sugars anaerobically leading to a build-up of lactic acid, causing our muscles to go into spasms (cramp) and fail eventually. The ensuing 'oxygen debt' means we need a prolonged period of recovery from exhaustion as we burn off the excess lactic acid, and the muscles build up their oxygen reserves again.

In short, just when we need it most, our respiratory system can fail and so it imposes a severe limitation on our abilities. Additionally, in order to shift enough oxygen and carbon dioxide across the alveolar membrane into and out of the blood, their lining needs to be very thin and is easily damaged leading to emphysema.

Birds, on the other hand, have a different system. Fresh air not only goes into their lungs on inspiration but into storage sacks too. When they breath out, fresh air from the storage sacks is pushed through the lungs, flushing out all the stale air, so a bird's lungs get fresh air in both phases of their respiration.

So, they are able to make do with finer tubes and more robust alveoli and can sustain prolonged effort with little muscle fatigue. In fact, the action of the wing muscles actually increases the rate of respiration without additional effort. Mammals, on the other hand get no special assistance from their locomotory system and need to bring in additional 'accessory' muscles to increase respiration when necessary, imposing yet another demand on the system.

Creationists believe these two systems were intelligently designed by the same designer who appears to have used the worst design for humans. Biologists, on the other hand, point out that evolution is utilitarian and makes do with whatever works, provided each improvement gives some advantage. They also point out that with an evolutionary system based on accumulated small changes over time, and which can't go in reverse, large-scale reorganisations are impossible, so branches in the evolutionary tree of life are often stuck with whatever worked for their ancestors because evolution can't and doesn't plan ahead. Instead evolution often consists of evolving work-arounds for inefficient earlier 'designs' so far as this is possible.

Again then, we have Creationism's putative designer showing it has the capability to design a far superior respiratory system for birds but opting to give humans and other mammals an inferior design. It hardly needs saying that the same putative designer used the mammalian respiratory system, with all its faults, for the marine species such as wales, seals, dugongs and manatees.

Sharks' Superior Immune System

It has been known for many years that sharks and rays (elasmobranchs) are highly efficient wound-healers and have a great resistance to cancers and infections by viruses and bacteria. To evolutionary biologists, this is probably the result of a very long evolutionary history as apical predators for over 400 million years, which has exposed them to all manner of pathogens and so high selective pressure for efficient genes.

Now research by a team from the Nova Southeastern University (NSU) Save Our Seas Shark Research Center, the Guy Harvey Research Institute (GHRI), and the Cornell University College of Veterinary Medicine has shown that this is probably due to unique modification of shark genes (119). Changes to two immune genes, legumain, and Bag1 were especially notable for their modification. They have counterparts in humans where overactivity has been associated with a whole range of cancers. In the sharks, these appear to have been highly modified.

The ability to heal rapidly and to resist any organisms that might be expected to infect open wounds is probably the result of sharks having a high proportion of immunity-related genes, and several that only express in elasmobranchs.

To an ID advocate, the evolutionary explanation is not available, of course, so the explanation must be that their putative intelligent designer decided to give the elasmobranchs immunity from almost everything, including cancers but decided not to give this superior immune system to humans and other animals to defend themselves from the nasty pathogens and cancers it had designed to make them sick and die.

Amphibians' Superior Regenerative Abilities.

No faith-healer ever regenerated an amputated limb and nor did a mammal ever grow one back spontaneously, with or without the benefit of prayers and magic spells. Certainly, there are claims which never turn out to be reliable and are never written up in the medical literature, and the official Catholic Church's miracle of the regeneration of Miguel

Juan Pellicer's leg in 17th Century rural Spain is a laughable example of an exposed scam and the gullibility of true believers (120). If it wasn't still a nice little earner for the Catholic Church in Calanda, near Zaragoza, Spain, the story would probably have been confined to the museum of religious frauds long ago and been forgotten by now.

The fact is, mammals do not regenerate whole limbs or other major body organs with or without divine intervention.

However, many amphibians do, and some reptiles can regenerate tails. Salamanders and newts are however the only vertebrates that can regenerate an entire limb. The best-known salamander with this capability is the axolotl, *Ambystoma mexicanum,* which spends most of its life in the larval form in which it is even capable of breeding. Axolotls are endangered in the wild but were one more common. They are also cannibalistic.

A hatching of axolotl eggs was followed by a frenzy of cannibalism ensuring a natural selection process in which the ones most capable of surviving were the ones that made it to sexual maturity and lived to breed again, passing their abilities on to their offspring. During this process, the loss of a limb to a sibling was a common occurrence, so the ability to survive this by growing a new limb was an obvious advantage. Of course, this evolutionary explanation is denied to Creationists for whom dogma dictates that they must attribute this ability to an intelligent designer and magic.

In fact, of course, humans, like all mammals **do** have very limited regenerative abilities. Broken bones heal by making new bone, muscle cells make new muscle cells to repair damage, liver cells make new liver cells and all tissues apart from brain and nerve routinely make new cells to replace old cells. Skin even regenerates to a degree following injury, providing the damage is not too extensive and enough skin cells survive to make new ones. Unlike in mammals, when an axolotl loses a limb or part of one, cells rapidly migrate to the damaged area to form a blastema which behaves like the limb bud of a developing embryo.

The problem that mammals have is that specialised cells have a fixed set of epigenetic settings that cannot be reversed, so the cells are not capable of producing anything other than cells of their own specialised type, so, no matter what cells migrate into a wound, they will never produce different types of cells.

However, salamanders like the axolotl have the same cell differentiation and specialisation brought about by epigenetics to turn off the unwanted genes, during their development as an embryo from an initial single-celled fertilised egg or zygote, as mammals do. The question then is how they manage to reset these epigenetic settings in the blastema to produce cells capable of making bone, muscles, nerves, blood vessels and skin.

Some light was shed on this by researchers from the DFG Research Center for Regenerative Therapies, Technische Universität Dresden, Germany (121), who used 'brainbow' axolotls whose cells had been labelled with florescent markers, making it possible to track individual cell types as they migrated into the blastema. They found that substances derived from platelets stimulated migration of different connective tissue cells into the blastema. In mammals, platelets in the blood release substances that initiate blood clotting. In axolotls, the equivalent cells initiate blastema formation. This is followed by waves of migration of cells of different types each of which contributes to the regrowth of the limb or part of the limb, depending partly on when they arrive in the blastema.

A later study by researchers from the Department of Evolutionary Genetics,, Max Planck Institute for Evolutionary Anthropology, the Research Institute of Molecular Pathology, Vienna, Center for Regenerative Therapies (CRTD) , Technische Universität Dresden, Germany and the Max Planck Institute of Molecular Cell Biology and Genetics, Dresden, Germany, showed that the adult cells that migrate into the blastema reset to a progenitor state from which they can then differentiate into different specialised cells (122).

In another experiment, pig retinal cells were shown to be able to regenerate new retinal cells when grown together with axolotl retinal cells but not on their own (123). The axolotl cells contain a substance to which the pig cells could respond. This shows that the ability to regenerate if something we lost rather than something axolotls and other salamanders have gained. Fossil evidence shows the common ancestors of modern salamanders also had regenerative abilities.

And of course, all developing mammalian embryos generate limbs from groups of undifferentiated cells in the limb buds!

The question for Creationists then is why did an intelligent designer decide to turn off this regenerative ability in mammals, including humans? Why does a supposedly omnibenevolent creator not want people to regrow lost limbs or even lost fingers?

Lastly, here is just one example of several were the 'design' of the human body looks more like a sub-optimal compromise and not the perfect design of a maximally good, omnipotent designer. It is based in part on a blog post I wrote following a holiday on the Côte d'Azur, France (124).

Rickets, Melanomas, Sunlight and Vitamin D.
Pale skin is generally recognised as an adaptive feature in Eurasian peoples because darker skin, which evolved in Africa, filtered out too much sun to make enough vitamin D - which is made in the skin in response to sunlight and we don't normally get enough in our diets.

Human beings need vitamin D for normal bone development in childhood. We can get it from two sources: from our food or by making it ourselves in our skin. Our diet is normally deficient in it so we need to supplement this by manufacturing it in our skin. This process involves ultraviolet-B (UVB) from the sun which converts a cholesterol-like substance known as provitamin D3 to vitamin D3

(125). This is then processed further by the liver and the kidneys to form vitamin D.

Deficiency in vitamin D in childhood leads to thin limb bones deficient in calcium phosphate which give the bones strength so the child may fail to grow to a normal height, may be susceptible to bone fractures and may well have deformed limbs, especially the lower leg bones which may be 'bowed'. Teeth may also fail to develop normally.

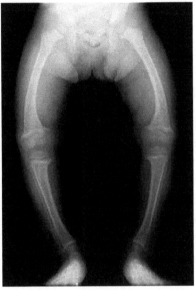

Figure 50 X-ray photo of child with rickets - Internet source

This syndrome is known as rickets and was a common developmental problem particularly in North-western Europe where it became especially prevalent after the Industrial Revolution in the northern industrial centres where poverty was rife, diet was poor, the climate was cloudy and the sun was weak even in summer. My father's step-father was crippled by rickets and could barely walk. Imagine what a handicap that would be in a hunter-gatherer!

In the UK where rickets was a problem, the provision of free milk at school as a post-war welfare measure was largely responsible for abolishing it.

But, with UVB comes ultraviolet-A (UVA) and too much exposure to UVA can cause cells in the skin to mutate and become cancerous, giving rise to melanoma, an especially aggressive cancer. There are also a couple of other skin cancers caused by too much exposure to direct sunlight (or sunbeds).

One of the reasons humans evolved dark skin as we lost body hair may well have been selective pressure caused by melanomas, in conditions where the sun was strong enough to produce enough vitamin D despite it being filtered by protective melanin in the skin.

When the immediate ancestors of modern human migrated out of Africa one of the problems preventing them extending their range northwards may well have been the incidence of rickets caused by lack of sunlight on a dark skin evolved in a part of the world where the seasons are not so well differentiated and where the sun is always high in the sky. It used to be thought that Europeans and northern Asians evolved pale skin by this selective pressure alone but it might well be that we acquired it from our Neanderthal cousins with whom there is now known to have been limited interbreeding (126).

Neanderthals had been evolving in Eurasia for some 200,000-250,000 years before *Homo sapiens* arrived on the scene and so had probably 'solved' many of the problems of living in colder, cloudier and more seasonal climates. In effect, we may have short-circuited evolution and acquired in a few thousand years what it had taken Neanderthals and their ancestors 250,000 years to evolve.

The problem is that the solution to the lack of UVB on our skins created the problem of too much UVA and so an evolutionary balancing act was created with Europeans walking a tightrope between rickets and melanomas. We mostly avoid rickets, though not very well in some situations and we mostly avoid melanomas, though not very well in some situations. It might well be that Neanderthals had evolved a defence to melanoma which, for some reason, we didn't inherit from them. We inherited half the package which was enough to allow us to move north but the trade-off was that we sometimes get melanoma and we sometimes have rickets.

In countries like Australia where pale-skinned people now live their lives in strong sunlight, melanoma is a significant problem. It is now increasing rapidly in the UK (from about 3% in 1975 to about 17% in 2011 (127)) since, beginning several decades ago, more people began travelling abroad to places like Spain, Greece and the Côte d'Azur to get a suntan on the beaches, or, as a fashion statement, began spending time on sunbeds to get a nice tan prior to going.

This is perfectly understandable and fully explained in general principles by a utilitarian evolutionary process which has no plan and no compassion, and fits perfectly into known human evolutionary history. It is completely inexplicable in terms of an intelligent design by a benign designer. In short, why would a compassionate designer design a problem which can only be solved by a balancing act between that problem and another one, neither of which benefit humans in any way? Did this intelligent, omniscient, omnibenevolent designer not realise his design was going to live in cloudy, northern climates and that getting enough vitamin D was going to be a problem? And did it not realise that its solution to this problem was going to cause another one?

Fortunately, science has not only explained the problem but has provided the cure - good diet and efficient sun-screening preparations. I use factor 30 because an anorak looks a bit silly on a beach and I remember all too well the blisters across my shoulders and on my shins from the second degree burns I got the first time I went abroad to bake on a Croatian beach in former Yugoslavia, 50 years ago.

This last example is something which is easy for evolutionary biology to explain in terms of the 'selfish' genes pushing evolution to produce something that ultimately benefits them but not the unfortunate carrier in whose body they find themselves. This is the cute little Australian marsupial mouse, *Antechinus stuartii,* also known as the brown antechinus.

The Marsupial Mouse.

The marsupial mouse, *A. stuartii,* is not a mouse or even closely related to mice. It is a marsupial and so more closely related to kangaroos, wombats and koalas, than the mice and shrews it superficially resembles. This similarity is the result of convergent evolution, where creatures with similar life-styles in similar environments tend to converge on similar body shapes.

Figure 51 Marsupial mouse, Antechinus stuartii

The reason this species is included here is because of what happens to the males as they look for a breeding partner: they literally die for sex.

The males spend so much time and energy looking for mates and mating that they all die at the end of the short breeding season due to the collapse of their immune systems caused by stress.

Unlike other small, placental mammals, females need to concentrate their efforts on rearing just one batch of young at a time when their insect food is at its peak. This means they have a very short period in which to mate and males need to concentrate all their reproductive efforts in this one short period. To make matters worse for the males, since the females' fertility is synchronized to that of the emergence of their food species, all the females in the area will be fertile for the same short period. Females are also promiscuous during short breeding period to ensure a successful breeding season.

In 2013, researchers led by Diana O. Fisher of Queensland University, School of Biological Sciences showed that this strategy has forced males into a 'suicidal' reproductive strategy of a frenzy of mating activity in which the corticosteroid feedback mechanism that normally limits stress, is overridden, causing a fatal collapse of their immune system (128).

Try as I might, I can think of no way that this can be presented as the work of an intelligent, all-loving creator deity or how that is a better explanation than the result of a mindless, unplanned and undirected natural process.

Clearly, adulation of whatever 'creator' could have created this and the very may examples I've just presented in this book is not the purpose of the Intelligent Design movement and Creationism. In the next chapter, I'll discuss what is going on here.

The Malevolent Designer

.

Conclusion

We have seen then that life needs life. Living things eat living things and maybe kill them in the process. Some are entirely dependent on others, often a particular species, for their continued survival. Viruses, things from the very edge of life – particles that cannot reproduce themselves without usurping some other species' machinery for replicating its DNA - do what they needed to do to maintain themselves, and the winners are those which produce the most viruses; the losers are those hosts who cannot live with the damage the virus does in the process. There is no plan, no emotion and no magic.

Life lives!

We have seen that living organisms are merely a resource in the environment of other organisms. Bacteria and single-celled protozoans which find themselves living in the bodies of other organisms evolve strategies to avoid being rejected and killed by the organism in self-defence, so we have evolutionary arms races – something no single intelligent designer would design since it is effectively competing with itself, designing ever more creative solutions to problems it created earlier; solutions which then become more problems to be solved. And, as we see, Creationism can only allow one creator because their dogma holds that it must comply fully with a literal interpretation of their favourite holy book – even if that leaves their putative creator looking like an incompetent fool.

If the control the dependent organism needs to exert involves taking control of the host organism to help ensure more of the parasites survive into the next generation, then natural selection will favour those strategies. So, we have the obscenities of *Anaplasma phagocytophilum* controlling the genes of the black-legged tick so it has more copies of the *rickettsia* organism to inject into its victims when it takes a blood meal and so make them sick. All to produce more copies of *A. phagocytophilum!*

We also have the hideous *Plasmodium falciparum* that destroys our red blood cells when we have malaria, having an inbuilt clock so they can co-ordinate their activity to make sure a feeding mosquito gets plenty of organisms to pass on when she next takes a blood meal, and having a special mechanism to cope with the defensive increase in body temperature their presence provokes in their victim, the better to produce more *P. falciparum* organisms - and make more people die from malaria.

Life lives!

If producing more copies of the chytrid fungi means millions of frogs and other amphibians die and 90 species go extinct, then so be it. There is nothing in evolution by natural selection which makes an emotional, moral judgement or which considers the consequences of a mutation. If that mutation, by whatever means, results in more copies of the chytrid fungus, then it will come to predominate in the *Batrachochytrium dendrobatidis* gene pool and dead frogs are mere collateral damage. If they can't play in the evolutionary arms race then they lose by default.

Life lives!

Our fleas, lice and bed bugs have all evolved to be the way they are and to live the way they do, because they evolved that way as our ancestors changed their environment and created new niches for them, so the random mutations took on new meanings in the context of these new environment – information that was given new meaning by the sieve of natural selection, iterated a hundred million times as we evolved firstly in Africa, then in Eurasia and the rest of the world. As we diversified from our common ancestor with the chimpanzees, so our lice diversified too, mapping their own evolution onto that of *Homo sapiens* and recording in their genome when we made the change from a naked ape on the African savannah to a clothed ape living in cooler northern climates and needing cloths to survive.

If the *Yersinia pestis* bacterium had a chance mutation that turned an endemic bacterium into a mass murderer, spread by rats and fleas to kill entire villages full of people and a couple of hundred million others, then that was an unfortunate, but entirely incidental and unplanned consequence of *Y. pestis* suddenly evolving a better way to make more copies of itself.

Life lives!

It matters not one bit to the genes of the *Fusarium oxysporum* fungus that banana growers lose their livelihood and humanity loses a valuable food resource because a mutation means they kill banana plants. All that matters is that the mutation now makes more *F. oxysporum* . It was not intended that way but evolution by natural selection ensured it happened that way.

So, we can go through all the nasty little parasitic worms, the avian brood parasites, the fish-eating lampreys, the vampire bats, the hideous snub-nosed eel living in the abdomen of a halibut, the cookiecutter shark tearing lumps out of seals and whales and the blood-sucking oxpeckers and say the same thing. Life lives and hosts continue to be the victims of parasites because that produces more parasites. The poor health, loss of control and death of the host is mere collateral damage; unplanned, unfortunate and unwanted; the consequence of a mindless natural process which had given us the biodiversity we see and a natural world to marvel at and ask questions of.

Life lives!

Nothing in nature dies a good death of old age. For almost all animals, life is nasty, brutish and short, laid low by parasites or killed by predators in the most painful way imaginable. Even if they live to old age, most will then die through being too old and weak to find food and water. Imagine if you can the death of a buffalo, suffocated by the killing bite of a lioness while others in the pride begin to disembowel you to get at your liver. Or the death of a gnu as it is drowned by a crocodile while having its limbs torn off. These sentient mammals will have the ability to feel pain which evolved to tell us something is wrong. Nothing in evolution is capable of designing a mechanism to turn pain off when it no longer serves a purpose. For these animals the end-of-life experience must be agonising and full of pain and fear. This should not be beyond the ability of a kindly, compassionate, merciful designer to solve, if it wanted to.

What we never see is evidence of direction by a compassionate, loving, merciful deity, determined to minimise the suffering in the world. A world with such evil in it could not possibly be the creation of a maximally good god. A maximally good god could not be ignorant of, indifferent to or incapable of preventing suffering. Nor could it be the architect of it.

Because of the manifest damage it is doing to religious belief by clinging tenaciously to absurd, easily-refuted ideas and with the intellectually bankrupt tactics it uses, Creationism's persistence is, as Francis S. Collins, former director of the Human Genome Project and a devout Christian, said in "The Language of God", is one of the great puzzles of our time.

> *Thus, by any reasonable standard, Young Earth Creationism has reached a point of intellectual bankruptcy, both in its science and in its theology. Its persistence is thus one of the great puzzles and great tragedies of our time.*

He went on to say:

> *By attacking the fundamentals of virtually every branch of science, [Young Earth Creationism] widens the chasm between the scientific and spiritual worldviews, just at a time where a pathway toward harmony is desperately needed. By sending a*

message to young people that science is dangerous, and that pursuing science may well mean rejecting religious faith, Young Earth Creationism may be depriving science of some of its most promising future talents.

But it is not science that suffers most here. Young Earth Creationism does even more damage to faith, by demanding that belief in God requires assent to fundamentally flawed claims about the natural world. Young people brought up in homes and churches that insist on Creationism sooner or later encounter the overwhelming scientific evidence in favor of an ancient universe and the relatedness of all living things through the process of evolution and natural selection. What a terrible and unnecessary choice they then face! To adhere to the faith of their childhood, they are required to reject a broad and rigorous body of scientific data, effectively committing intellectual suicide. Presented with no other alternative than Creationism, is it any wonder that many of these young people turn away from faith, concluding that they simply cannot believe in a God who would ask them to reject what science has so compellingly taught us about the natural world. (129 pp. 176-177)

But is Collins right that the survival of religious fundamentalism of which Creationism is but a manifestation, such a great puzzle? Writing in *Psychology Today*, the psychologist, David Ludden Ph.D., professor of psychology at Georgia Gwinnett College, said:

*For people who are high in a personality trait called **social dominance orientation**, society is naturally organized as a hierarchy. On this view, which rung on the social ladder you belong to is determined by obvious identifiers such as gender and race...*

*A common characteristic of people high in **social dominance orientation** is self-enhancement, which is the tendency to see*

*yourself in an overly favorable manner. Such people feel
driven to engage in acts that support their superiority or
protect their delicate but overinflated egos.*

*For example, although many such people are quite ignorant of
history, science, or even the basic facts of the world, they're
unwilling to admit what they don't know. As a result, they're
highly susceptible to the "false facts" of government
propaganda and entertainment media masquerading as news
networks. They also tend to engage in overclaiming, that is,
asserting that they know certain patently false concepts to be
true* (130).

That last paragraph will probably strike a chord with anyone who has
tried to engage Creationists in the social media and come across
someone who is manifestly ignorant of biology and basic science but
nevertheless regards him/herself as a leading expert in the subject, fully
qualified to tell the experts that they've got it all wrong.

Professor Ludden was commenting on the findings by three researchers,
led by Daniel N. Jones of the University of Nevada, that:

*Agentic self-enhancement consists of self-protective and self-
advancing tendencies that can lead to aggression, especially
when challenged. Because self-enhancers often endorse
aggression to defend or enhance the self-concept, religious self-
enhancement should lead to endorsing aggression to defend or
enhance one's religion* (131).

Fundamentalist religion, in addition to gender and race, is an important
social identifier too. It seems it inculcates people with a hierarchical
view of society where social dominance is king, probably reinforced by
hierarchies within their brand of fundamentalism, and where they see
themselves as low down in the social order. Creationism then, in their
own eyes, and in the eyes of other Creationists, is a vehicle for
attempted self-enhancement, where self-proclaimed expertise and being

privy to secret knowledge and understanding is assumed to elevate them above the ordinary. The aggression and sometimes overt threats that any risk of exposure of their ignorance produces, is a self-defence mechanism to the perceived threat of exposure of this illusory 'self-enhancement' as phony and so worthless.

There are probably few better ways to conclude a book refuting the notion that the natural world is the intended design of an intelligent, compassionate, omnibenevolent, maximally good god than by repeating Judge John E. Jones, who said in his judgement in Kitzmiller vs Dover District School Board, rejecting Intelligent design as science and finding that it was Christian fundamentalist Creationism posing as science:

> We find that ID fails on three different levels, any one of which is sufficient to preclude a determination that ID is science. They are: (1) ID violates the centuries-old ground rules of science by invoking and permitting supernatural causation; (2) the argument of irreducible complexity, central to ID, employs the same flawed and illogical contrived dualism that doomed creation science in the 1980's; and (3) ID's negative attacks on evolution have been refuted by the scientific community. As we will discuss in more detail below, it is additionally important to note that ID has failed to gain acceptance in the scientific community, it has not generated peer-reviewed publications, nor has it been the subject of testing and research. (132 p. 65)

Possibly, the most damning finding was that several fundamentalist Christian members of the Dover District School Board, having sworn an oath on the Bible to tell the truth, lied about their religious beliefs and their previous advocacy of Intelligent design.

> Finally, although Buckingham, Bonsell, and other defense witnesses denied the reports in the news media and contradicted the great weight of the evidence about what transpired at the June 2004 Board meetings, the record reflects

that these witnesses either testified inconsistently, or lied outright under oath on several occasions, and are accordingly not credible on these points. (132 p. 105)

Christian fundamentalists lied under oath to try to trick a court to allow them to lie to children in science class! I will leave the reader to draw their own conclusions from this dishonesty and what it says of the political motives behind intelligent design Creationism.

When you show the world you know you need to lie for your faith, you show the world you know your faith is a lie that requires people to believe falsehoods. What good is there in a religion that requires its followers to be deceived?

What possible good reason could there be to deceive people into believing falsehoods, especially about what you believe to be your god's creation?

Figure 52 Elephant Calf

Bibliography

1. **Rubicondior, Rosa.** *The Unintelligent Designer: Exposing the Intelligent Design Hoax.* Abingdon : CreateSpace, 2018. ISBN: 978-1723144219.

2. *Creationism and conspiracism share a common teleological bias.* **Pascal, Wagner-Egger, et al.** 16, s.l. : Elsevier Inc., 20 August 2018, Current Biology, Vol. 28, pp. R867-R868. ISSN 09609822.

3. *Judgments About Fact and Fiction by Children From Religious and Nonreligious Backgrounds.* **Corriveau, Kathleen H., Chen, Eva E. and Harris, Paul L.** 2, s.l. : John Wiley & Sons, Inc., 1 March 2015, Cognitive Science, Vol. 39, pp. 353-382. Free access.

4. *Zombie soldier beetles: Epizootics in the goldenrod soldier beetle, Chauliognathus pensylvanicus (Coleoptera: Cantharidae) caused by Eryniopsis lampyridarum (Entomophthoromycotina: Entomophthoraceae).* **Steinkraus, Donaldc C., Hajek, A and Liebherr, J.** s.l. : Elsevier Inc, 2017, Journal of Invertebrate Pathology, Vol. 148, pp. 51-59.

5. *The first Laboulbeniales (Ascomycota, Laboulbeniomycetes) from an American millipede, discovered through social media.* **Santamaria, Sergi, Enghoff, Henrik and Reboleira, Ana Sofia.** 14 May 2020, MycoKeys, Vol. 67, pp. 45-53.

6. *Chronic bee paralysis as a serious emerging threat to honey bees.* **Budge, Giles E., et al.** 2164, s.l. : Springer Nature, 1 May 2020, Nature Communications, Vol. 11. Open access.

7. **Moate, Maddie.** What would happen if bees went extinct? *BBC Future.* [Online] [Cited: 21 July 2020.]

https://www.bbc.com/future/article/20140502-what-if-bees-went-extinct.

8. **Center For Disease Control and Prevention.** CDC Center For Disease Control and Prevention - Anaplasmosis. *Signs and Symptoms.* [Online] 11 January 2019. [Cited: 21 July 2020.] https://www.cdc.gov/anaplasmosis/symptoms/index.html.

9. *Repression of tick microRNA-133 induces organic anion transporting polypeptide expression critical for Anaplasma phagocytophilum survival in the vector and transmission to the vertebrate host.* **Ramasamy, Ellango, et al.** 16, 2 July 2020, PLOS Genetics, Vol. 7, p. e1008856.

10. **Miller, Kenneth R.** *Only a Theory: Evolution and the Battle for America's Soul.* New York : Penguin Books, 2008. ISBN-10: 0143115669.

11. **Behe, Michael J.** *Darwin's Black Box: The Biochemical Challenge to Evolution.* 2nd Rev. s.l. : Free Press, 2006. ISBN-10: 0743290313.

12. **National Center For Science Education.** Kitzmiller v. Dover: Intelligent Design on Trial. *NCSE National Center for Science Education.* [Online] 15 December 2015. [Cited: 21 July 2020.] https://ncse.ngo/kitzmiller-v-dover-intelligent-design-trial.

13. *Campylobacter jejuni motility integrates specialized cell shape, flagellar filament, and motor, to coordinate action of its opposed flagella.* **Cohen, E.J., et al.** 16, 2 July 2020, PLOS Pathology, Vol. 7, p. e1008620.

14. *The minimal meningococcal ProQ protein has an intrinsic capacity for structure-based global RNA recognition.* **Bauriedl, Saskia, et al.** 2823, s.l. : Springer Nature, 4 June 2020, Nature Communications, Vol. 11.

15. *Grad-seq guides the discovery of ProQ as a major small RNA-binding protein.* **Smirnov, Alexandre, et al.** 41, s.l. : National Academy of Science, 11 October 2016, PNAS, Vol. 114, pp. 11591-11596;.

16. *The Major RNA-Binding Protein ProQ Impacts Virulence Gene Expression in Salmonella enterica Serovar Typhimurium.* **Westermann, Alexander J., et al.** 1, s.l. : American Society for Microbiology, January 2019, mBio, Vol. 10, pp. e02504-18.

17. *Extraordinary Flux in Sex Ratio.* **Charlat, Sylvain, et al.** 5835, s.l. : American Association for the Advancement of Science, 13 July 2007, Science, Vol. 317, p. 214.

18. *Evolving MRSA: High-level β-lactam resistance in Staphylococcus aureus is associated with RNA Polymerase alterations and fine tuning of gene expression.* **Panchal, Viralkumar V., et al.** [ed.] Andreas Peschel. 7, s.l. : PLOS, 24 July 2020, PLOS Pathogens, Vol. 16, p. e1008672. Open access.

19. **Center for Disease Control and Prevention.** Salmonella Homepage. *CDC.* [Online] US Government, 5 December 2019. [Cited: 1 August 2020.] https://www.cdc.gov/salmonella/general/index.html.

20. *Identification of Novel Mobilized Colistin Resistance Gene mcr-9 in a Multidrug-Resistant, Colistin-Susceptible Salmonella enterica Serotype Typhimurium Isolate.* **Carroll, Laura M., et al.** [ed.] Mark S. Turner. 3, s.l. : American Society for Microbiology, 25 June 2019, mBiol, Vol. 10.

21. **Center for Disease Control and Prevention.** Parasites - Toxoplasmosis (Toxoplasma infection). *CDC Center for Disease Control and Prevention.* [Online] 5 September 2018. [Cited: 22 July 2020.] https://www.cdc.gov/parasites/toxoplasmosis/disease.html.

22. *Toxoplasma gondii Co-opts the Unfolded Protein Response To Enhance Migration and Dissemination of Infected Host Cells.* **Augusto, Leonardo, et al.** [ed.] Louis M. Weiss. 4, s.l. : American Society for Microbiology, 25 August 2020, mBio, Vol. 11, pp. e00915-20. Open access.

23. *Rats, cats, people and parasites: the impact of latent toxoplasmosis on behaviour.* **Webster, Joanne P.** 12, s.l. : Elsevier B.V., October 2001, Microbes and Infection, Vol. 3, pp. 1037-1045.

24. *Fatal attraction in rats infected with Toxoplasma gondii.* **Berdoy, M., Webster, J. P. and Macdonald, D. W.** 1452, s.l. : The Royal Society, 7 August 2000, Procedings of the Royal Society B, Vol. 267, pp. 1591-1594.

25. *Morbid attraction to leopard urine in Toxoplasma-infected chimpanzees.* **Poirotte, Clémence, et al.** 3, s.l. : Cell Press, 8 February 2016, Current Biology, Vol. 26, pp. R98-R99.

26. *Neuraminidase antigenic drift of H3N2 clade 3c.2a viruses alters virus replication, enzymatic activity and inhibitory antibody binding.* **Powell, Harrison and Pekosz, Andrew.** 6, s.l. : PLOS, 29 June 2020, PLOS Pathologt, Vol. 16, p. e1008411.

27. *Spatial mapping of polymicrobial communities reveals a precise biogeography associated with human dental caries.* **Kim, Dongyeop, et al.** 22, s.l. : National Academy of Science of the United States of America, 2 June 2020, PNAS, Vol. 117, pp. 12375-12386.

28. *Mucosal microbiome dysbiosis in gastric carcinogenesis.* **Coker, Olabisi Oluwabukola, et al.** 6, s.l. : BMJ Journals, 01 06 2018, Gut, Vol. 67, pp. 1024-1032.

29. *Periodontal disease, tooth loss, and risk of oesophageal and gastric adenocarcinoma: a prospective study.* **Lo, Chun-Han, et al.** s.l. : BMJ Journal, 30 June 2020, Gut, Vol. 2020. Letter.

Bibliography

30. *Mixed Red-Complex Bacterial Infection in Periodontitis.* **Salo, Tuula, et al.** s.l. : Hindawi Publishing Corporation, 6 March 2013, International Journal of Dentistry, Vol. 2013, pp. 1687-8728.

31. *Phosphatidylinositol 3-phosphate and Hsp70 protect Plasmodium falciparum from heat-induced cell death.* **Lu, Kuan-Yi , et al.** s.l. : eLifeSciences, 25 September 2020, eLife, Vol. 9, p. e56773. Open access.

32. **Miller, Kenneth R.** Edging Toward Irrelevence. *http://www.millerandlevine.com/.* [Online] December 2014. [Cited: 22 July 2020.] http://www.millerandlevine.com/evolution/behe-2014/Behe-1.html.

33. **Behe, Michael J.** *The Edge of Evolution.* Reprinted. New York : The Free Press, 2008. p. 336. ISBN-10: 0743296222.

34. **Rubicondior, Rosa.** An 'Intelligent Design' That Really Gets Up Your Nose. *Rosa Rubicondior.* [Online] 15 May 2015. [Cited: 6 August 2020.] https://rosarubicondior.blogspot.com/2015/05/an-intelligent-design-that-really-gets.html.

35. *Pathogenesis of amoebic encephalitis: Are the amoebae being credited to an 'inside job' done by the host immune response?* **Baig, Abdul Mannan.** s.l. : Elsevir Science Direct, 1 August 2015, Acta Tropica, Vol. 148, pp. 72-76.

36. **NHS.** Chlamydia. *NHS.* [Online] UK National Health Service, 4 June 2018. [Cited: 4 August 2020.] https://www.nhs.uk/conditions/chlamydia/.

37. *Reprogramming of host glutamine metabolism during Chlamydia trachomatis infection and its key role in peptidoglycan synthesis.* **Rajeeve, Karthika, et al.** s.l. : Springer Nature, 3 August 2020, Nature Microbiology.

163

38. **Center for Disease Control and Prevention.** Zika Virus. *Zika Virus -Health Effects & Risks.* [Online] 20 November 2019. [Cited: 22 July 2020.] https://www.cdc.gov/zika/healtheffects/index.html.

39. *Structural basis for STAT2 suppression by flavivirus NS5.* **Wang, Boxiao, et al.** s.l. : Springer Nature Ltd., 10 August 2020, Nature Structural & Molecular Biology. 1545-9985.

40. *Zika virus infection enhances future risk of severe dengue disease.* **Katzelnick, Leah C., et al.** 6507, s.l. : American Association for the Advancement of Science, 28 August 2020, Science, Vol. 369, pp. 1123-1128. 1095-9203.

41. *Increased growth ability and pathogenicity of American- and Pacific-subtype Zika virus (ZIKV) strains compared with a Southeast Asian-subtype ZIKV strain.* **Kawai, Yasuhiro, et al.** 13, 6 June 2009, PLOS Neglected Tropical Diseases, Vol. 6, p. e0007387.

42. *Amphibian fungal panzootic causes catastrophic and ongoing loss of biodiversity.* **Scheele, Ben C., et al.** 6434, s.l. : American Association for the Advancement of Science (AAAS), 29 March 2019, Science, Vol. 363, pp. 1459-1463.

43. *Sustained immune activation is associated with susceptibility to the amphibian chytrid fungus.* **Savage, Anna E., et al.** s.l. : John Wiley & Sons, Inc, 23 July 2020, Molecular Ecology, pp. 1-15.

44. *Behavioral betrayal: How select fungal parasites enlist living insects to do their bidding.* **Lovett, Brian, et al.** [ed.] Donald C. Sheppard. 6, s.l. : PLOS, 18 June 2020, PLOS Pathogens, Vol. 16, p. e1008598.

45. *Psychoactive plant- and mushroom-associated alkaloids from two behavior modifying cicada pathogens.* **Boyce, Greg R., et al.** s.l. : Elsevier Inc., 2 June 2019, Fungal Ecology, Vol. 41, pp. 147-164.

46. **Johns Hopkins University.** New Cases. *Corona Virus Resource Center.* [Online] 9 October 2020. [Cited: 9 October 2020.] https://coronavirus.jhu.edu/.

47. **Guardian Newspaper.** Covid vaccine tracker: when will a coronavirus vaccine be ready? *The Guardian.* [Online] 23 September 2020. [Cited: 23 September 2020.] Guardian Covid Vaccine Tracker. https://www.theguardian.com/world/ng-interactive/2020/oct/07/covid-vaccine-tracker-when-will-a-coronavirus-vaccine-be-ready.

48. **Sauer, Lauren.** Johns Hopkins Medicine - Health. *What Is Coronavirus?* [Online] 22 July 2020. [Cited: 23 July 2020.] https://www.hopkinsmedicine.org/health/conditions-and-diseases/coronavirus.

49. *Enhanced Binding of SARS-CoV-2 Spike Protein to Receptor by Distal Polybasic Cleavage Sites.* **Baufo, Qiao and Olvera de Cruz, Monica.** s.l. : American Chemical Society, 2 August 2020, ACS Nano. 1936-0851.

50. **Stony Brook University.** Genomic Basis of Bat Superpowers Revealed: Like How They Survive Deadly Viruses. *Stony Brook University News.* [Online] 22 July 2020. [Cited: 3 August 2020.] https://news.stonybrook.edu/newsroom/genomic-basis-of-bat-superpowers-revealed-like-how-they-survive-deadly-viruses/.

51. *Six reference-quality genomes reveal evolution of bat adaptations.* **Jebb, David, et al.** 7817, s.l. : Springer Nature, 1 July 2020, Nature, Vol. 583, pp. 578–584. Open access.

52. *Tracking Changes in SARS-CoV-2 Spike: Evidence that D614G Increases Infectivity of the COVID-19 Virus.* **Korber, B., et al.** s.l. : Elsevier Inc., 2 July 2020, Cell Press. Elsevier Inc..

53. **National Institute of Allergy and Infectious Diseases (NIH).** NIH Diseases - Conditions. *Coronaviruses.* [Online] 19 May 2020. [Cited:

23 July 2020.] https://www.niaid.nih.gov/diseases-conditions/coronaviruses.

54. *Type I and III interferons disrupt lung epithelial repair during recovery from viral infection.* **Major, Jack, et al.** s.l. : American Association for the Advancement of Science, 11 June 2020, Science, p. eabc2061. Online.

55. *SARS-CoV-2 Infectivity and Neurological Targets in the Brain.* **Lukiw, Walter J., Pogue, Aileen and Hill, James M.** s.l. : Springer Nature, 25 August 2020, Cellular and Molecular Neurobiology. 1573-6830.

56. *Broad host range of SARS-CoV-2 predicted by comparative and structural analysis of ACE2 in vertebrates.* **Damas, Joana, et al.** s.l. : National Academy of Science of the United States of American , 21 August 2020, Proceedings of the National Academy of Sciences, p. 202010146. Open access. 1091-6490.

57. **Willingham, Emily.** Scientific American. *Of lice and men: An itchy history.* [Online] 14 February 2011. [Cited: 24 July 2020.] https://blogs.scientificamerican.com/guest-blog/of-lice-and-men-an-itchy-history/.

58. *Pair of lice lost or parasites regained: the evolutionary history of anthropoid primate lice.* **Reed, David L., et al.** 7, s.l. : Springer Nature, 7 March 2007, BMC Biology, Vol. 5.

59. *What's in a name: The taxonomic status of human head and body lice.* **Light, Jessica E., Toups, Melissa A. and Reed, David L.** 3, s.l. : Elsevier Inc., June 2008, Molecular Phylogenetics and Evolution, Vol. 47, pp. 1203-1216.

60. **NHS.** NHS. *Typhus.* [Online] 2 October 2017. [Cited: 24 July 2020.] https://www.nhs.uk/conditions/typhus/.

61. **Los Angeles Times.** California's first plague case in 5 years is confirmed in South Lake Tahoe. *Los Angeles Times.* [Online] 27 August 2020. [Cited: 18 August 2020.] https://www.latimes.com/california/story/2020-08-17/californias-first-plague-case-in-5-years-is-confirmed-in-south-lake-tahoe.

62. *Early Divergent Strains of Yersinia pestis in Eurasia 5,000 Years Ago.* **Rasmussen, Simon, et al.** 3, s.l. : Elsevier Inc, 22 October 2015, Cell Press, Vol. 163.

63. **Rubicondior, Rosa.** Malevolent Designer News - Making Better Mosquitoes. *Rosa Rubicondior.* [Online] 29 April 2019. [Cited: 2 August 2020.] https://rosarubicondior.blogspot.com/2019/04/malevolent-designer-news-making-better_23.html.

64. *Aedes aegypti Mosquitoes Detect Acidic Volatiles Found in Human Odor Using the IR8a Pathway.* **Raji, oshua I., et al.** 8, s.l. : Cell Press, 28 March 2019, Current Biology, Vol. 29, pp. 1253-1262.e7. Open access.

65. *Lycaenid Caterpillar Secretions Manipulate Attendant Ant Behavior.* **Hojo, Masaru K., Pierce, Naomi E. and Tsuji, Kazuki.** 17, s.l. : Elsevier Inc., 31 August 2015, Current Biology, Vol. 25, pp. 2260-2264.

66. **Michener, Charles D.** *The Bees of the World.* Second. Baltimore : Johns Hopkins University Press, 2007. p. 992. ISBN-10: 0801885736.

67. **Falk, Stephen and Lewington, Richard.** *Field Guide to the Bees of Great Britain and Ireland.* Reissue. London & New York : Bloomsbury Wildlife - Bloomsbury Publishing PLC., 2018. p. 432. ISBN-10: 1472967054.

68. *Parasitoid Increases Survival of Its Pupae by Inducing Hosts to Fight Predators.* **Grosman, Amir H., et al.** [ed.] Nigel E. Raine. 6, s.l. : PLOS, 3 July 2008, PLOS One, Vol. 3, p. e2276. Open access.

69. *Recurrent Domestication by Lepidoptera of Genes from Their Parasites Mediated by Bracoviruses.* **Gasmi, Laila, et al.** 9, s.l. : PLOS, 17 September 2015, PLOS Genetics, Vol. 11, p. e1005470.

70. *Host manipulation by an ichneumonid spider ectoparasitoid that takes advantage of preprogrammed web-building behaviour for its cocoon protection.* **Takasuka, Keizo, et al.** 15, s.l. : The Company of Biologists Ltd., 1 August 2015, The Journal of Experimental Biology, Vol. 218, pp. 2326-2332. 1477-9145.

71. *Tongue Replacement in a Marine Fish (Lutjanus guttatus) by a Parasitic Isopod (Crustacea: Isopoda).* **Brusca, Richard C. and Gilligan, Matthew R.** 3, s.l. : American Society of Ichthyologists and Herpetologists (ASIH), 16 August 1983, Copeia, Vol. 1983, pp. 813-816.

72. *New records of fish-parasitic isopods (Cymothoidae) in the Eastern Pacific (Galápagos and Costa Rica).* **Williams, Ernest H. and Bunkley-Williams, Lucy.** December 2003, NOTICIAS DE GALÁPAGOS, Vol. 62, pp. 21-23.

73. **Piper, Ross.** *Extraordinary Animals: An Encyclopedia of Curious and Unusual Animals.* Westport, Connecticut : Greenwood Press, 2007. p. 320. ISBN-10: 0313339228.

74. *An Unprecedented Role Reversal: Ground Beetle Larvae (Coleoptera: Carabidae) Lure Amphibians and Prey upon Them.* **Wizen, Gil and Gasith, Avital.** [ed.] Sean A. Rands. 9, s.l. : PLoS, 21 September 2011, PLoS ONE, Vol. 6, p. e25161. Open access.

75. **Smithsonian.** Smithsonian BugInfo. *Beetles.* [Online] [Cited: 28 August 2020.] https://www.si.edu/spotlight/buginfo/beetle.

76. *Homage to Linnaeus: How many parasites? How many hosts?* **Dobson, Andy, et al.** Supplement 1, s.l. : National Accademy of Science of the United States of America, 12 August 2008, Proceedings of the National Academy of Sciences (PNAS), Vol. 105, pp. 1482-11489.

77. *Regulation of the host immune system by helminth parasites.* **Maizels, Rick M. and McSorley, Henry J.** s.l. : Elsevier Inc., 29 July 2016, The Journal of Allergy and Clinical Immunology, pp. 666–675.

78. *Maternal filarial infection as risk factor for infection in children.* **Lammie, P.J., et al.** 8748, s.l. : The Lancet, 27 April 1991, The Lancet, Vol. 337, pp. 1005-1006.

79. *Effect of helminth-induced immunity on infections with microbial pathogens.* **Salgame, P., Yap, G. and Gause, W.** s.l. : Nature, 23 October 2013, Nature Immunology, Vol. 14, pp. 1118–1126.

80. *Behavioural manipulation in a grasshopper harbouring hairworm: a proteomics approach.* **Biron, D.G., et al.** 1577, s.l. : The Royal Society Publishing, 31 August 2005, Procedings of the Royal Society B, Vol. 272, pp. 2117–2126. Open access.

81. *Myrmeconema neotropicum n. g., n. sp., a new tetradonematid nematode parasitising South American populations of Cephalotes atratus (Hymenoptera: Formicidae), with the discovery of an apparent parasite-induced host morph.* **Poinar, George and Yanoviak, Stephen P.** 2, s.l. : Springer Nature, 1 February 2008, Systematic Parasitology, Vol. 69, pp. pages145–153.

82. *Do Leucochloridium sporocysts manipulate the behaviour.* **Weslowska, W and Weslowski, T.** s.l. : The Zoological Society of London, 28 October 2013, Journal of Zoology, Vol. 292, pp. 151-155.

83. *When parasites disagree: Evidence for parasite-induced sabotage of host manipulation.* **Hafer, N and Melinski, M.** 3, s.l. : Wiley Online, 31 January 2015, Evolution, Vol. 69, pp. 611-620. Open access.

84. *Behavioural Responses to Simulated Avian Predation in Female Three Spined Sticklebacks: The Effect of Experimental Schistocephalus Solidus Infections.* **Barber, Iain, Svensson, P. Andreas and Walker, Peter.** 11-12, 1 January 2004, Behaviour, Vol. 141, pp. 1425–1440.

85. **Roberts, Larry and Janovy, John.** *Foundations of Parasitology.* s.l. : McGraw-Hill Education , 2008. ISBN-10: 0071311033.

86. **World Health Organization (WHO).** Dracunculiasis (guinea-worm disease). *WHO Factsheet.* [Online] 16 March 2020. [Cited: 29 July 2020.] https://www.who.int/en/news-room/fact-sheets/detail/dracunculiasis-(guinea-worm-disease).

87. *From Parasitism to Mutualism: Unexpected Interactions Between a Cuckoo and Its Host.* **Canestrari, Daniela, et al.** 6177, s.l. : American Association for the Advancement of Science, 21 March 2014, Science, Vol. 343, pp. 1350-1352.

88. **Rubicondior, Rosa.** Evolutionists Have Even More to Crow About. *Rosa Rubicondior.* [Online] 21 March 2014. [Cited: 29 July 2020.] https://rosarubicondior.blogspot.com/2014/03/evolutionists-have-even-more-to-crow.html.

89. *How to evade a coevolving brood parasite: egg discrimination versus egg variability as host defences.* **Spottiswoode, Claire N. and Stevens, Martin.** 1724, s.l. : Royal Society Publishing, 13 April 2011, Proceedings of the Royal Society B: Biological Sciences, Vol. 278, pp. 3566-3573.

90. *Multimodal mimicry of hosts in a radiation of parasitic finches.* **Jamie, Gabriel A., et al.** s.l. : Wiley Online, 21 July 2020, Evolution. Open access. 0014-3820.

91. **Le Cointe, Paul.** *L'Amazonie brésilienne. Le Pays, ses habitants, ses ressources. Notes et statistiques jusqu'en 1920.* Paris : A. Challamel, , 1922.

92. **Rubicondior, Rosa.** A New Angle on Sex for Creationists. *Rosa Rubicondior.* [Online] Rosa Rubicondior, 7 February 2012. [Cited: 3 August 2020.] https://rosarubicondior.blogspot.com/2012/02/new-angle-on-sex-for-creationists.html.

93. *The immunogenetics of sexual parasitism.* **Swann, Jeremy B., et al.** s.l. : American Association for the Advancement of Science, 30 July 2020, Science, p. eaaz9445. Published online.

94. *The sea lamprey germline genome provides insights into programmed genome rearrangement and vertebrate evolution.* **Smith, Jeramiah J., et al.** 2, s.l. : Springer Nature, 22 January 2018, Nature Genetics, Vol. 50, pp. 270–277. Open access.

95. *The allopatric phase of speciation: the sharp-beaked ground finch (Geospiza difficilis) on the Galápagos islands.* **Grant, Peter R., Grant, B. Rosemary and Petren, Kenneth.** 3, s.l. : The Linnean Socity of London , 14 January 2008, Biological Journal of the Linnean Society, Vol. 69, pp. 287-317.

96. **Fullerton Baird, Spencer.** *The Sea Fisheries Of Eastern North America.* s.l. : Sagwan Press, 2015. p. 232. ISBN-10: 1340053322.

97. *First Documented Attack on a Live Human by a Cookiecutter Shark (Squaliformes, Dalatiidae: Isistius sp.).* **Honebrink, Randy, et al.** s.l. : University of Hawaii, Honolulu, July 2011, Pacific Science, pp. 365-374.

98. **Various.** *Handbook of the Birds of the World: Bush-Shrikes to Old World Sparrows.* [ed.] Josep Del Hoyo, Andy Elliott and David Christie. s.l. : Lynx Edicions, 2009. p. 894. Vol. 14. ISBN-10: 8496553507.

99. *Host-parasite tissue adhesion by a secreted type of β-1,4-glucanase in the parasitic plant Phtheirospermum japonicum.* **Kurotani, Kenichi, et al.** 1, s.l. : Springer Nature, 30 July 2020, Communications Biology, Vol. 3, p. 407. Open access. 2399-3642.

100. *Complete Plastid Genome of the Recent Holoparasite Lathraea squamaria Reveals Earliest Stages of Plastome Reduction in Orobanchaceae.* **Samigullin, Tahir H., et al.** [ed.] Marc Robinson-Rechavi. 3, 2 March 2016, PLOS One, Vol. 11, p. e0150718.

101. **Centre for Agriculture and Bioscience International (CABI).** Invasive Species Compendium - Datasheet 37747. *Orobanche ramosa (branched broomrape).* [Online] Centre for Agriculture and Bioscience International, 19 November 2019. [Cited: 1 August 2020.] https://www.cabi.org/isc/datasheet/37747.

102. **Shaw, Jonathan.** Colossal Blossom: Pursuing the peculiar genetics of a parasitic plant. *Harvard Magazine.* [Online] Harvard Magazine Inc., 2017. [Cited: 1 August 2020.] https://harvardmagazine.com/2017/03/colossal-blossom.

103. *Possible Loss of the Chloroplast Genome in the Parasitic Flowering Plant Rafflesia lagascae (Rafflesiaceae).* **Molina, Jeanmaire, et al.** 4, s.l. : Oxford Academic, 23 January 2014, Molecular Biology and Evolution, Vol. 31, pp. 793-803.

104. *A specialized avian seed dispersal system in a dry-fruited nonphotosynthetic plant, Balanophora yakushimensis.* **Suetsugu, Kenji.** [ed.] John Pastor. s.l. : The Ecological Society of America, 26 July 2020, Ecology, p. e03129. Online Version of Record before inclusion in an issue. 0012-9658.

105. *Volatile Chemical Cues Guide Host Location and Host Selection by Parasitic Plants.* **Runyon, Justin B., Mescher, Mark C. and De Moraes, Consuelo M.** 5795, 29 June 2006, Science, Vol. 313, pp. 1964-1967.

Bibliography

106. *Cuscuta australis (dodder) parasite eavesdrops on the host plants' FT signals to flower.* **Shen, Guojing, et al.** [ed.] David C. Baulcombe. s.l. : National Academy of Sciences of the United States of America, 28 August 2020, Proceedings of the National Academy of Sciences, p. 202009445. 1091-6490.

107. *Plant defenses against parasitic plants show similarities to those induced by herbivores and pathogens.* **Runyon, Justin B., Mescher, Mark C. and De Moraes, Consuelo M.** 8, s.l. : Taylor & Francis online, 1 August 2010, Plant Signaling & Behavior, Vol. 5, pp. 929-931.

108. **Rubicondior, Rosa.** Evolution in Progress - Eurasian Blackcap. *Rosa Rubicondior.* [Online] 31 January 2019. [Cited: 1 August 2020.] https://rosarubicondior.blogspot.com/2019/01/evolution-in-progress-eurasian-blackcap.html.

109. *Haustorial Structure and Functioning of the Root Hemiparastic Tree Nuytsia floribunda (Labill.) R.Br. and Water Relationships with its Hosts.* **Calladine, Ainsley and Pate, John S.** 6, s.l. : Oxford Academic, 1 June 2000, Annals of Botany, Vol. 85, pp. 723–731.

110. *Botanical parasitism of an insect by a parasitic plant.* **Egan, Scott P., et al.** 16, s.l. : Elsevier Ltd, 20 August 2018, Current Biology, Vol. 28, pp. R863-R864.

111. **Boyd, Jade.** Love vine sucks life from wasps, leaving only mummies. *Rice University News and Media Relations.* [Online] Rice University, 20 August 2018. [Cited: 3 September 2020.] http://news.rice.edu/2018/08/20/love-vine-sucks-life-from-wasps-leaving-only-mummies/.

112. **Rice University, [perf.].** *Rice U. biologists document dueling parasites on South Florida oak trees.* [Video]. Rice University; YouTube, 2018.

173

113. *Identification of pollen allelochemical in Hieracium × dutillyanum Lepage and its ecological impacts on Conyza canadensis (L.) Cron. and Sonchus arvensis L. dominated community in southern Ontario, Canada.* **Murphy, Stephen D., et al.** 1, 2009, Allelopathy Journal, Vol. 23, pp. 85-94.

114. **Wikipedia.** Peto's paradox. *Wikipedia.* [Online] Wikipedia, 8 June 2020. [Cited: 6 August 2020.] https://en.wikipedia.org/wiki/Peto%27s_paradox.

115. *TP53 copy number expansion is associated with the evolution of increased body size and an enhanced DNA damage response in elephants.* **Sulak, Michael, et al.** e11994, s.l. : eLife Sciences Publications, Ltd, 19 September 2016, eLife, Vol. 5.

116. *Wild hummingbirds discriminate nonspectral colors.* **Stoddard, Mary Caswell, et al.** 26, s.l. : National Academy of Sciences of the United Staes of America, 30 June 2020, Proceedings of the National Academy of Sciences (PNAS), Vol. 117, pp. 15112-15122.

117. **Fuller-Wright, Liz.** Wild hummingbirds see a broad range of colors humans can only imagine. *Princeton University News.* [Online] Princeton University Communications Department, 15 June 2020. [Cited: 7 August 2020.] https://www.princeton.edu/news/2020/06/15/wild-hummingbirds-see-broad-range-colors-humans-can-only-imagine.

118. **Rubicondior, Rosa.** More Blunders By The Unintelligent Designer. *Rosa Rubicondior.* [Online] 4 April 2014. [Cited: 6 August 2020.] https://rosarubicondior.blogspot.com/2014/04/more-blunders-by-unintelligent-designer.html.

119. *Comparative transcriptomics of elasmobranchs and teleosts highlight important processes in adaptive immunity and regional endothermy.* **Marra, Nicholas J., et al.** 1, s.l. : Springer Nature, 30 January 2017, BMC Genomics, Vol. 18, p. 87. Open access.

120. **Rubicondior, Rosa.** The Mircle of Miracles. *Rosa Rubicondior.* [Online] 14 February 2013. [Cited: 7 August 2020.] https://rosarubicondior.blogspot.com/2013/02/the-miracle-of-miracles.html.

121. *Live Imaging of Axolotl Digit Regeneration Reveals Spatiotemporal Choreography of Diverse Connective Tissue Progenitor Pools.* **Currie, Joshua D., et al.** 4, s.l. : Elsevir BV Science Direct, 21 November 2016, Developmental Cell, Vol. 39, pp. 411-423. Open access.

122. *Single-cell analysis uncovers convergence of cell identities during axolotl limb regeneration.* **Gerber, Tobias, et al.** 6413, s.l. : American Association for the Advancement of Science., 26 October 2018, Science, Vol. 362, p. eaaq0681.

123. **Arntsen, Emily.** The salamander that eats its siblings' arms could one day help you grow a new one. *Phys Org.* [Online] Phys.org, 25 October 2019. [Cited: 8 August 2020.] Provided by Northeastern University. https://phys.org/news/2019-10-salamander-siblings-arms-day.html.

124. **Rubicondior, Rosa.** Malignant Melanoma - Another Gift From The Intelligent Designer? *Rosa Rubicondior.* [Online] 7 September 2014. [Cited: 27 August 2020.] https://rosarubicondior.blogspot.com/2014/09/malignant-melanoma-another-gift-from.html.

125. **Matteson, Cade.** How does skin produce vitamins? *How Stuff Works - Health.* [Online] HowStuffWorks.com, 20 August 2009. [Cited: 27 August 2020.] https://health.howstuffworks.com/skin-care/information/anatomy/skin-produce-vitamins.htm.

126. **Rubicondior, Rosa.** Rosa Rubicondior. *So What Did The Neanderthals Ever Do For Us?* [Online] 5 May 2014. [Cited: 27

August 2020.] http://rosarubicondior.blogspot.com/2014/05/so-what-did-neanderthals-ever-do-for-us.html.

127. **Cancer Research UK.** Melanoma skin cancer incidence statistics. *Cancer Research UK.* [Online] Cancer Research UK. [Cited: 27 August 2020.] https://www.cancerresearchuk.org/health-professional/cancer-statistics/statistics-by-cancer-type/melanoma-skin-cancer/incidence#trends.

128. *Sperm competition drives the evolution of suicidal reproduction in mammals.* **Fisher, Diana O., et al.** [ed.] James H. Brown. s.l. : National Academy of Sciences of the United States of America, 2 October 2013, Proceedings of the National Academy of Sciences, p. 201310691. Online. 1091-6490.

129. **Collins, Francis S.** *The Language of God: A Scientist Presents Evidence for Belief.* New Ed. London : Simon & Schuster UK Ltd., 2007. p. 304. Paperback. ISBN-10: 1847390927.

130. **Ludden, David.** Why People Kill in the Name of God: The role of self-enhancement in religious aggression. *Psychology Today.* [Online] Psychology Today, 8 June 2020. [Cited: 2 August 2020.] https://www.psychologytoday.com/us/blog/talking-apes/202006/why-people-kill-in-the-name-god.

131. *Religious Overclaiming and Support for Religious Aggression.* **Jones, Daniel N., et al.** s.l. : Sage Journals, 14 April 2020, Social Psychological and Personality Science.

132. **United States District Court for the Middle District of Pennsylvania.** Memorandum of opinion - Judge John E. Jones. *Kitzmiller v. Dover: Intelligent Design on Trial.* [Online] 20 December 2005. [Cited: 31 July 2020.] https://ncse.ngo/files/pub/legal/kitzmiller/highlights/2005-12-20_Kitzmiller_decision.pdf.

133. **Johns Hopkins University.** New Cases of COVID-19 In World Countries. *Johns Hopkins Coronavirus Resource Center.* [Online] 23 July 2020. [Cited: 23 July 2020.] Updated daily. https://coronavirus.jhu.edu/data/new-cases.

134. *Identifying airborne transmission as the dominant route for the spread of COVID-19.* **Zhang, Renyi, et al.** 26, s.l. : National Academy of Sciences of the Ubited Staes of America, 30 June 2020, PNAS, Vol. 117, pp. 14857-14863.

135. **Conger, Kate, Healy, Jack and Tompkins, Lucy.** Churches Were Eager to Reopen. Now They Are Confronting Coronavirus Cases. *New York Times - Morning Briefing.* [Online] 10 July 2020. [Cited: 23 July 2020.] https://www.nytimes.com/2020/07/08/us/coronavirus-churches-outbreaks.html.

136. **Times of Israel.** Israeli rabbi: Coronavirus outbreak is divine punishment for gay pride parades. *The Times of Israel.* [Online] 8 March 2020. [Cited: 23 July 2020.] https://www.timesofisrael.com/israeli-rabbi-blames-coronavirus-outbreak-on-gay-pride-parades/.

137. **Daily Telegraph.** South London faith healer promises his followers Covid-19 protection with oil and red yarn. *Daily Telegraph.* [Online] 1 April 2020. [Cited: 23 July 2020.] https://www.telegraph.co.uk/news/2020/04/01/south-london-faith-healer-promises-followers-covid-19-protection/.

138. **Office for National Statistics.** Coronavirus (COVID-19) related deaths by religious group, England and Wales: 2 March to 15 May 2020. *5.Religious group differences in deaths involving COVID-19, adjusted for socio-demographic factors.* [Online] 19 June 2020. [Cited: 23 July 2020.] https://www.ons.gov.uk/peoplepopulationandcommunity/birthsdeathsandmarriages/deaths/articles/coronaviruscovid19relateddeathsbyreligiousgroupenglandandwales/2marchto15may2020.

139. **Woodward, Alex.** The Independent. *'A phantom plague':*
America's Bible Belt played down the pandemic and even cashed in.
Now dozens of pastors are dead. [Online] 24 April 2020. [Cited: 23
July 2020.] https://www.independent.co.uk/news/world/americas/bible-
belt-us-coronavirus-pandemic-pastors-church-a9481226.html.

140. *Rapid real-time tracking of non-pharmaceutical interventions and*
their association with SARS-CoV-2 positivity: The COVID-19 Pandemic
Pulse Study. **Clipman, Steven J, et al.** s.l. : Oxford University Press , 2
September 2020, Clinical Infectious Diseases, p. ciaa1313. Online.
1058-4838.

Index

Index

Index

Index

Figure 53 Head of a Peregrine Falcon, Falco peregrinus

List of Illustrations

Other Books by Rosa Rubicondior

The Light of Reason Series:

The Light of Reason: And Other Atheist Writings.
Irreverent essays, thought-provoking articles and humorous items on atheism, religion, science, evolution, creationism and related issues.

    (Paperback) ISBN-10: 1516906888, ISBN-13: 978-1516906888   £9.95 (US $14.95)
    (Kindle) ASIN: B014N0IPVI   £3.95 (US $5.99)

The Light of Reason: Volume II – Atheism, Science and Evolution.
Thought-provoking essays on the conflict between fundamentalist religion and science, and exposing the anti-science, extremist political agenda of the modern creationist industry.

    (Paperback) ISBN-10: 1517105188, ISBN-13: 978-1517105181   £9.95 (US $14.95)
    (Kindle) ASIN: B014N0IR16   £3.99 (US $5.99)

The Light of Reason: Volume III – Apologetics, Fallacies, and Other Frauds.
Thought-provoking essays and articles on religion and atheism, dealing with religious apologetics, fallacies, miracles and other frauds

    (Paperback) ISBN-10: 151710761X, ISBN-13: 978-1517107611   £6.95 (US $9.95)
    (Kindle) ASIN: B014N0IRE8   £2.99 (US $3.99)

The Light of Reason: Volume IV - The Silly Bible.
Exposing the absurdities, contradictions and historical inaccuracies in the Bible and advancing the case for atheism and against religion. This volume, the fourth in the Light of Reason series, deals with contradictions and absurdities in the Bible.

    (Paperback) ISBN-10: 1517108209, ISBN-13: 978-1517108205   £8.95 (US $13.95)
    (Kindle) ASIN: B014N0IR8E   £3.99 (US $4.99)

The Light of Reason: And Other Atheist Writing. (all 4 volumes in one e-book)
Based on the Rosa Rubicondior science and Atheism blog, this is a collection of Atheist and science articles, some short, others lengthier, exploring the interface between religion and science and which have been published over some four years.

    (Kindle only) ASIN: B013DYOK32   £6.34 (US $9.95)

Other books on science, Atheism and theology

An Unprejudiced Mind: Atheism, Science & Reason.
Essays on science and theology from a scientific atheist perspective, exploring particularly evolution versus creationism.

    (Paperback) ISBN-10: 1522925805, ISBN-13: 978-1522925804   £9.95 (US $14.95)
    (Kindle) ASIN: B019UGXPM4   £3.99 (US $5.95)

The Malevolent Designer

Ten Reasons To Lose Faith: And Why You Are Better Off Without It.
Why faith is not only a fallacy and useless as a route to the truth but is actually harmful to society and to the individual. It systematically dismantles the standard religious apologetics and shows them to be bogus and deliberately constructed to mislead.

(Paperback) ISBN-13:978-1530431953, ISBN–10: 1530431956 £10.75 (US $14.75)
(Kindle) ASIN: B01DGVO3JS   £6.95 (US $8.95)

What Makes You So Special? From the Big Bang to You.
How did you come to be here, now? This book takes you from the Big Bang to the evolution of modern humans and the history of human cultures.

(Paperback) ISBN-13: 978-1546788294, ISBN-10: 1546788298 £8.95 (US $11.50)
(Kindle) ASIN: B071FTKXLZ   £6.20 (US $8.25)

The Unintelligent Designer: Refuting the Intelligent Design Hoax.
The evidence against intelligent design in nature. This book presents evidence of stupidity in arms races, of prolific waste and needless complexity compared to the principles of good design - minimal waste, maximal simplicity and clear purpose.

(Paperback) ISBN-13: 978-1723144219; ISBN-10: 1723144215   £8.45 (US $10.75)
(Kindle) ASIN: B07G121BMK   £4.55 (US $5.95)

The Internet Handbooks series

The Internet Creationists' Handbook: A Joke for the Rest of Us.
A humorous look at creationist apologetics on the Internet, showing the fallacies and dishonest tactics creationists are using to try to recruit scientifically illiterate people into their political cult.

(Paperback) ISBN-13: 978-1721605149, ISBN-10: 1721605149 £5.25 (US $7.50)
(Kindle) ASIN: B07DZF75KD   £3.75 (US $5.00)

The Christian Apologists' Handbook: A Joke for the Rest of Us.
A humorous look at Christian apologetics on the Internet, showing the fallacies and dishonest tactics Christian fundamentalists are using to try to recruit scientifically and theologically illiterate people to their cults, often with political motives.

(Paperback) ISBN-13: 978-1721724727, ISBN–10: 1721724729   £5.25 (US $7.50)
(Kindle) ASIN: B07DYDVMW4   £3.75 (US $5.00)

The Muslim Apologists' Handbook: A Joke for the Rest of Us.
A humorous look at Muslim apologetics on the Internet, showing the fallacies and dishonest tactics Muslim fundamentalists are using to try to recruit scientifically and theologically illiterate people to their cuts, often with political motives.

(Paperback) ISBN-13: 978-1721756896, ISBN-10: 1721756892   £5.25 (US $7.50)
(Kindle) ASIN: B07DZF75KD $3.75 (US $5.00)

Lightning Source UK Ltd.
Milton Keynes UK
UKHW021823200722
406135UK00009B/1769